ALLE ZEIT WACH 1842

Springer Series in Synergetics

Synergetics, an interdisciplinary field of research, is concerned with the cooperation of individual parts of a system that produces macroscopic spatial, temporal or functional structures. It deals with deterministic as well as stochastic processes.

Volume 1 **Synergetics** An Introduction 2nd Edition
By H. Haken

Volume 2 **Synergetics** A Workshop
Editor: H. Haken

Volume 3 **Synergetics** Far from Equilibrium
Editors: A. Pacault and C. Vidal

Volume 4 **Structural Stability in Physics**
Editors: W. Güttinger and H. Eikemeier

Volume 5 **Pattern Formation** by Dynamic Systems and **Pattern Recognition**
Editor: H. Haken

Volume 6 Dynamics of **Synergetic Systems**
Editor: H. Haken

Volume 7 Problems of **Biological Physics**
By L. A. Blumenfeld

Volume 8 **Stochastic Nonlinear Systems** in Physics, Chemistry, and Biology
Editors: L. Arnold and R. Lefever

Volume 9 Numerical Methods in the **Study of Critical Phenomena**
Editors: J. Della Dora, J. Demongeot and B. Lacolle

Numerical Methods in the Study of Critical Phenomena

Proceedings of a Colloquium, Carry-le-Rouet
France, June 2–4, 1980

Editors:
J. Della Dora, J. Demongeot, and B. Lacolle

With 83 Figures

Springer-Verlag Berlin Heidelberg New York 1981

Dr. Jean Della Dora
Dr. Jacques Demongeot
Dr. Bernard Lacolle

Université Scientifique et Médicale de Grenoble
Laboratoire d'Informatique et de Mathématiques Appliquées de Grenoble
F - BP 53 X 38041 Grenoble Cédex, France

Volume Editor:

Professor Dr. Hermann Haken

Institut für Theoretische Physik der Universität Stuttgart,
Pfaffenwaldring 57/IV,
D-7000 Stuttgart 80, Fed. Rep. of Germany

ISBN-13: 978-3-642-81705-2 e-ISBN-13: 978-3-642-81703-8
DOI: 10.1007/978-3-642-81703-8

Softcover reprint of the hardcover 1st edition 1981

2153/3130-543210

Preface

This volume contains most of the lectures presented at the meeting held in Carry-le-Rouet from the 2nd to the 4th June 1980 and entitled "Numerical Methods in the Study of Critical Phenomena".

Scientific subjects are becoming increasingly differentiated, and the number of journals and meetings devoted to them is continually increasing. Thus it has become very difficult for the non-specialist to approach subjects with which he is not familiar. Hence the purpose of our meeting was to bring together scientists from different disciplines to study a common subject and to stimulate discussion between participants. We hope this goal was reached.

The lectures are grouped in five chapters and, inside the first and the second chapter, under two headings. In each group they are classified in alphabetical order by author.

We are pleased to publish these Proceedings in a series whose multidisciplinary character has been emphasized from the beginning.

We are indebted to all who provided us with their help, particularly to Mrs. A. Litman of the Centre International de Rencontres Mathématiques at Luminy (C.I.R.M.) whose kindness and efficiency are well known; from the practical point of view, the meetings were organized within the scientific framework of the G.I.S. No.19 (C.N.R.S.), with the participation of the University of Grenoble.

Grenoble, June 1981

J. Della Dora
J. Demongeot
B. Lacolle

Contents

Chapter 1. *Mathematical Methods* 1

1.1 Study of Singularities

Padé-Hermite Approximants. By J. Della Dora 3

Somme Comments About the Numerical Utilization of Factorial Series
By J.P. Ramis and J. Thomann 12

Introduction to Real Quasianalytic Classes and Continuation Problems
By T. Reboul 26

1.2 Critical Phenomena in Dynamical Systems

Groups Transformations and Critical Asymptotics Applications to Non-Linear Differential and Partial Derivative Equations. By J.R. Burgan, M.R. Feix, E. Fijalkow, M.P. Moraux, and A. Munier 37

Antecedent Invariant Curves of an Endomorphism. Influence Domain of a Stable Cycle Coexisting with an Isolated Stable Invariant Curve
By R.L. Clerc and Ch. Hartmann 47

Topological Entropy As a Measure of Dynamic Chaos in Endomorphisms
By J. Couot, C. Gillot, I. Gumowski, and C. Mira 54

Topological Entropy of Markov Processes for a C^0-Endomorphism of the Interval. By C. Gillot and G. Gillot 57

Sequential Iteration of Threshold Functions. By E. Golès Chacc 64

Some Properties of Second Order Dynamic Systems with Parametric Resonances
By I. Gumowski 71

Chapter 2. *Applications in Physics* 77

2.1 Critical Phenomena in Solid-State Physics

On the Bifurcation of Certain Kam Tori in the Standard Mapping
By S. Aubry 79

MO Stochasticity Criterion. By R. Caboz and A. Lonke 81

Singularities in Saw Numerical Simulations. By F. Cuozzo and E.E. Cambiaggio 86

Monte Carlo Measurement of the Single Vortex Free Energy in the Kosterlitz-Thouless Theory. By H.J. Hilhorst, H.N.J. Vogelij, C. van Leeuwen, and B.P.Th. Veltman 97

Algebraic Method for the Computation of the Partition Functions of Spin Glasses and Numerical Study of the Distributions of Zeros. By B. Lacolle 104

Percolation and Gelation by Additive Polymerization. By P. Manneville and L. de Sèze 116

Ground State Structure of the Random Frustration Model in Two Dimensions By R. Maynard and R. Rammal 125

Line Defects and the Glass Transition. By N. Rivier and D.M. Duffy 132

Universality in Size-Effects in 2D Percolation. By J. Rousseng and H. Ottavi 143

2.2 Use of Renormalisation Techniques

The Phenomenological Renormalization Method. By B. Derrida and J. Vannimenus 153

Computation of the Yang-Lee Edge Singularity in Ising Models By P. Moussa 159

Real-Space Renormalization-Group Method for Quantum Systems: Application to Quantum Frustration in Two Dimensions. By K.A. Penson, R. Jullien, P. Pfeuty, and K. Uzelac 166

Yang-Lee Edge Singularity by Real Space Renormalization Group By K. Uzelac, R. Jullien, P. Pfeuty, and P. Moussa 171

Chapter 3. *Applications in Biology* 177

Numerical Determination of a Periodical Solution of Discontinuous Type, near a Singular Point, for a Neurophysiological Model. By J. Argemi and B. Rossetto 179

On the Relation Between the Logical Structure of Systems and Their Ability to Generate Multiple Steady States or Sustained Oscillations By R. Thomas 180

Critical Delays in Logical Asynchronous Models. By A. Verhamme and P. Van Ham 194

Chapter 4. *Applications in Chemistry* 201

A Simulation Technique for Studying Critical Properties of Chemical Dissipative Systems. By P. Hanusse 203

Critical Paths and Passes: Application to Quantum Chemistry
By D. Liotard and J.-P. Penot 213

Chapter 5. *Non-Physical Applications of Statistical Mechanics* 223

Telephone Network: Statistical Mechanics and Non-Random Connecting
Procedures. By E. Bonomi, J.L. Lutton, and M.R. Feix 225

The Thermodynamic Formalism in Population Biology. By L. Demetrius 233

Asymptotic Inference for Markov Random Fields on Z^d. By J. Demongeot 254

List of Contributors 269

Chapter 1
Mathematical Methods

1.1 Study of Singularities

Padé-Hermite Approximants

J. Della Dora

Université de Grenoble I, Laboratoire d'Informatique et de Mathématiques Appliquées, BP : 53 X, F-38041 Grenoble Cédex, France

1. Introduction to Padé-Hermite Forms ([1],[2],[3])

The result will be given for three formal power series. Nevertheless the theory is general; we can work with any number of series.

k will be a commutative field (in the application $k = \mathbb{R}$ or $\mathbb{C}$) and $k[[X]]$ the algebra of formal power series on k in one unknown.

Definition : Let f_1, f_2, f_3 be three formal power series belonging to $k[[X]]$.

Let $\bar{n} = (n_1, n_2, n_3)$ be a triple of positive integers ($n_i \in \mathbb{N}$).
We call Padé-Hermite (P-H) form of order $\bar{n}$ associated to (f_1, f_2, f_3) *every* triple of polynomials

$$(Y_1^{\bar{n}}, Y_2^{\bar{n}}, Y_3^{\bar{n}})$$

such that

- •) $Y_i^{\bar{n}}$ belongs to $k[X]$
- •) $\partial^\circ Y_i^{\bar{n}} \leq n_i \qquad (i=1,2,3)$
- •) $\sum_{i=1}^{3} Y_i^{\bar{n}} f_i = O(X^{|\bar{n}|+2})$

$(|\bar{n}| = n_1 + n_2 + n_3)$.

Remarks :

- •) In this context we can take into account the main interpolatory approximants wich are frequently used (Padé [4], D-Log and Gammel differential approximants [5], Shafer-Quadratic approximants [6]).
- •) A P-H form always exists.
- •) There is an infinity of P-H forms associated to the same $\bar{n}$. So we have a difficult normalisation problem.

This last remark leads us to defined *normalised* P-H forms .

For this purpose we introduced the three polynomials

$$
Z_i^{\bar{n}}(X) = \det \left|
\begin{array}{cc|cc|cc}
f_0^1 & & f_0^i & & f_0^3 & \\
 & \ddots & & \ddots & & \ddots \\
 & f_0^1 & & f_0^i & & f_0^3 \\
\vdots & \vdots & \vdots & \vdots & \vdots & \vdots \\
f^1_{|\bar{n}|+1} & f^1_{|\bar{n}|-n_1+1} & f^i_{|\bar{n}|+1} & f^i_{|\bar{n}|-n_i+1} & f^3_{|\bar{n}|+1} & f^3_{|\bar{n}|-n_3+1} \\
0 & 0 & 1 & X \ldots X^{n_i} & 0 & 0
\end{array}
\right|
$$

$(i=1,2,3 \; ; \; f_i(X) = \sum_{j=0}^{+\infty} f_j^i X^j)$

and the determinants $\Delta_{\bar{n}}^j$ which are obtained there when we fill the last row of $Z_i^{\bar{n}}$ by the following row :

$$(f^1_{|\bar{n}|+3+j}, \ldots, f^1_{|\bar{n}|+3+j-n_1} \; ; \; \ldots \; ; \; f^3_{|\bar{n}|+3+j}, \ldots, f^3_{|\bar{n}|+3+j-n_3}) \; .$$

With this notation we get the identity

$$\sum_{i=1}^{3} Z_i^{\bar{n}}(X) \, f_i(X) = X^{|\bar{n}|+2} \left(\sum_{j=0}^{+\infty} \Delta_{\bar{n}}^{j-i} \, X^j \right) \; .$$

We call this triple $(Z_1^{\bar{n}}, Z_2^{\bar{n}}, Z_3^{\bar{n}})$ *the normalised form* of order $\bar{n}$ associated to (f_1, f_2, f_3) .

We must take care about the following disagreements

1) we can have $Z_i^{\bar{n}} \equiv 0$ $(i=1,2,3)$

2) if the $Z_i^{\bar{n}}$ are not coprime there are very difficult problems (for more information see [1]) .

2. The Δ and P-H Tables

With the Δ determinants we can build tables which look like the Padé and C-tables of Gragg ([7],[8]). For this construction we use some identities wich are the direct generalization of the classical ones.

2.1. The Generalized Frobenius Formula

Let $\bar{n} = (n_1,n_2,n_3)$ and e_i $(i=1,2,3)$ the canonical basis vectors of $\mathbb{R}^3$.
We also note $\check{n} = \bar{n} - \sum_{i=1}^{3} e_i$. Then we can prove

$$\Delta^{-1}_{\bar{n}}\ \Delta^{-1}_{\check{n}} = \sum_{i=1}^{3} (-1)^{i+1}\ \Delta^{1}_{\check{n}+e_i}\ \Delta^{-1}_{\bar{n}-e_i} .$$

The proof is not an easy trick, see [2] or [3] for the general case and for a certain number of related formula.

2.2. The Sylvester Formulas

If α,p,q,r are elements of $\mathbb{N} \cup \{-1\}$ we prove the following relations :

$$\Delta^{\alpha}_{p,q,r}\ \Delta^{-1}_{p,q-1,r-1} = \Delta^{\alpha+1}_{p,q-1,r}\ \Delta^{-1}_{p,q,r-1} - \Delta^{\alpha+1}_{p,q,r-1}\ \Delta^{-1}_{p,q-1,r}$$

$$\Delta^{\alpha}_{p,q,r}\ \Delta^{-1}_{p-1,q-1,r} = \Delta^{\alpha+1}_{p-1,q,r}\ \Delta^{-1}_{p,q-1,r} - \Delta^{\alpha+1}_{p,q-1,r}\ \Delta^{-1}_{p-1,q,r}$$

$$\Delta^{\alpha}_{p,q,r}\ \Delta^{-1}_{p-1,q,r-1} = \Delta^{\alpha+1}_{p-1,q,r}\ \Delta^{-1}_{p,q,r-1} - \Delta^{\alpha+1}_{p,q,r-1}\ \Delta^{-1}_{p-1,q,r} .$$

Where $\alpha \geq -1$, $p,q,r \geq -1$ or 0 according to $p-1,q-1$ or $r-1$ appear in the formula.

With these two kinds of formulas we can build a family of algorithms such that if we know

$$f^i_0,\dots,f^i_N \quad ; \quad i=1,2,3$$

we can form tables filled with

- •) $\Delta^{\alpha}_{p,q,r}$, $(p+q+r+\alpha \leq N-3)$
- •) or triples $(Z^{\bar{n}}_i)_{i=1,2,3}$.

The reader must pay attention to the fact that we build in [1] or [2] algorithms fitting with triples $(f_i)_{i=1,2,3}$ such that $\Delta^{-1}_{\bar{n}} \neq 0$ for every $\bar{n}$ in $(\mathbb{N} \cup \{-1\})^3$.

In the general case we must take care of the Δ^{-1}-table block structure.

3. Zeros Block Structure in the Δ^{-1} Table

The interesting case appears when we find zeros in one of the following three positions :

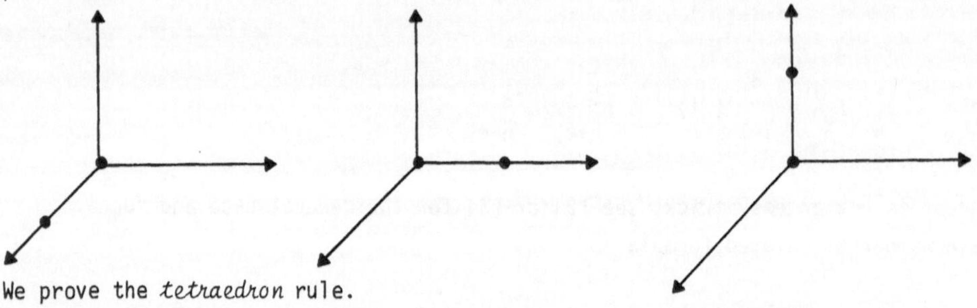

We prove the *tetraedron* rule.

Proposition : If $\Delta^{-1}_{p,q,r} = 0$ and $\Delta^{-1}_{p,q,r+1} = 0$ then under the following hypothesis

$$\Delta^{-1}_{p-1,q,r} \cdot \Delta^{-1}_{p,q-1,r} \cdot \Delta^{-1}_{p,q,r-1} \neq 0$$

$$\Delta^{-1}_{p-1,q,r+1} \cdot \Delta^{-1}_{p,q-1,r+1} \neq 0$$

we can prove that

$$\Delta^{0}_{p,q,r} = 0 \quad ; \quad \Delta^{-1}_{p+1,q,r} = 0 \quad ; \quad \Delta^{-1}_{p,q+1,r} = 0 \quad .$$

More generally we have the following representation (Figure 1).

This rule has a reciprocal :

Proposition : If $\Delta^{-1}_{p,q,r} = 0$ and if $\Delta^{-1}_{p',q',r'} \neq 0$ for (p',q',r') such that (p,q,r) belongs to the ℓ^1-neighbourhood of (p',q',r') then

$$\Delta^{0}_{p,q,r} = 0 \text{ involves } \Delta^{-1}_{p+1,q,r} = \Delta^{-1}_{p+1,q,r} = \Delta^{-1}_{p,q,r+1} = 0 \ .$$

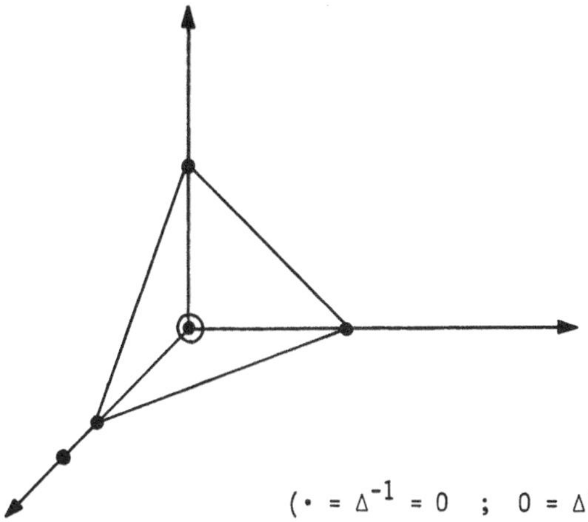

Fig. 1. $(\cdot = \Delta^{-1} = 0 \quad ; \quad 0 = \Delta^{0} = 0)$

The consequences are :

1) Suppose we have (for example)

$\Delta^{-1}_{p,q,r} = \Delta^{-1}_{p,q,r+1} = \Delta^{-1}_{p,q,r+2} = 0$ and that (for example)

$$\begin{cases} \Delta^{-1}_{p-1,q,r} \cdot \Delta^{-1}_{p,q-1,r} = \Delta^{-1}_{p,q,r-1} \neq 0 \\ \Delta^{-1}_{p-1,q,r-1} \cdot \Delta^{-1}_{p,q-1,r+1} \neq 0 \\ \Delta^{-1}_{p-1,q,r+2} \cdot \Delta^{-1}_{p,q-1,r+2} = 0 , \end{cases}$$

we can prove that

$$\Delta^{0}_{p,q+1,r} = \Delta^{-1}_{p,q+2,r} = \Delta^{-1}_{p+1,q+1,r} = 0$$

$$\Delta^{0}_{p+1,q,r} = \Delta^{-1}_{p+1,q,r} = 0$$

This involves $\Delta^{0}_{p,q,r} = \Delta^{0}_{p,q+1,r} = \Delta^{0}_{p+1,q,r} = \Delta^{0}_{p,q,r+1} = 0$ and this implies

$\Delta^{1}_{p,q,r} = 0$.

Now the reader can play very easily with the zeros of the Δ-tables.

Apart from the tetraedron rule we must have some ideas about the existence of unbounded blocks in the Δ^{-1}-table ; apart from the pathological cases related to the nullity of some elements in the begining of the series a case is very interesting :

We recall that three series belonging to $k[[X]]$ are said to be independant over $k(X)$ (the fraction field of the algebra of polynomials $k[X]$) if we cannot find 3 rational fractions such that

$$\sum_{i=1}^{3} P_i(X)\, f_i(X) \equiv 0 \quad .$$

(Indeed $P_i \not\equiv 0$; $i=1,2,3$) . It is obvious that we can only use $P_i \in k[X]$. Then we have :

Theorem : (f_1,f_2,f_3) are dependant over $k(X)$ if and only if there exists $\bar{n} = (n_1,n_2,n_3)$ such that for every $j \geq -1$ we have $\Delta^j_{\bar{n}} = 0$ and if $\bar{n}$ is in generic position .

This theorem has an interesting consequence :

If we have at the origin of $\mathbb{C}$ a branch (holomorphic in zero) of an algebraic function. If we do not know the equation of this algebraic function we can build the Δ^{-1}-table associated to the triple

$(f^2,f,1)$

and look for the point $\bar{n}$ such that $\Delta^j_{\bar{n}} = 0$ for every j .

Then using the $(Z_i^{\bar{n}})$-algorithm we restore the algebraic equation and the other branches of the function.

At the end we see that $\Delta^j_{\bar{n}} = 0$ for every j involves (using the tetraedron rule) the existence of an unbounded block of zeros :

$$\Delta^j_{\bar{m}} = 0 \quad , \quad \forall\, j \geq -1 \quad , \quad \forall\, \bar{m} \text{ such that } \bar{m} \geq \bar{n}$$

(here we use the lexicographical order).

4. Some Convergence Results (only for Algebraic Approximants)

Let Ω be a connected open set of $\mathbb{C}$, $\mathcal{H}(\Omega)$ the field of meromorphic functions over Ω and f a quadratic element belonging to $\mathcal{H}(\Omega)$.
That is to say f satisfies an equation

(1) $\quad f^2 + \gamma_1 f + \gamma_2 \equiv 0$

on Ω . Where γ_1 and γ_2 are two elements of $\mathcal{H}(\Omega)$ and (1) is irreducible.

We also need the following hypothesis :

1. There is an open disk

   $DO(0,\rho_0) = \{z \; ; \; z \in \mathbb{C} \text{ and } |z| < \rho_0\}$ enclosed in Ω , such that γ_1 and γ_2 have neither pole nor zero on

   $\partial DO(0,\rho_0) = \{z \; ; \; z \in \mathbb{C} \text{ and } |z| = \rho_0\}$.

2. We write $\gamma_i(X) = \dfrac{\hat{\phi}_i(X)}{\prod_{j=1}^{n_i} (X-x_j^i)}$

   $(x_j^i \in DO(0,\rho_0)$; $\hat{\phi}_i$ holomorphic in $DO(0,\rho_0)$ such that

   .) $\hat{\phi}_i\,(z) \neq 0 \qquad (z \in \partial\, DO(0,\rho_0))$

   .) $\hat{\phi}_i(x_j^i) \neq 0 \qquad (\forall\; j=1,\dots,n_i)$.

Under these conditions we write (1)

$$q(X)\; f^2 + \theta_1(X)f + \theta_2(X) = 0$$

with $q(X) = \prod_{j=1}^{n_1} (X-x_j) \prod_{j=1}^{n_2} (X-x_j^2)$, θ_1 and θ_2 holomorphic in $DO(0,\rho_0)$.

3. f admits an holomorphic branch at zero, which has the development

   $$f(X) = \sum_{j=0}^{+\infty} a_j\, X^j \; .$$

We know that the singularities of f belong to

$$\mathcal{D}(f,\mathcal{D}0(0,\rho_0)) = \{a \; ; \; a \in \mathcal{D}0(0,\rho_0) \; ; \; q(a)\; (\theta_1^2(a)-4q(a)\theta_2(a)) = 0\} \; .$$

Now we build the normalised P-H forms associated with $(f^2,1)$.
Let the roots of the equation

$$Z_1^{\bar{n}}(X)\; Y^2 + Z_2^{\bar{n}}(X)Y + Z_3^{\bar{n}}(X) \;=\; 0 \quad , \text{ be } h_1^{\bar{n}} \text{ and } h_2^{\bar{n}} \; .$$

It is natural to call $h_i^{\bar{n}}$ the approximants of f. With all this information we can prove :

<u>Proposition</u> : If 0 does not belong to $\mathcal{D}(\bar{n}) = \{z \; ; \; z \in \mathcal{D}0(0,\rho_o) \; ;$

$Z_1^{\bar{n}}\,((Z_2^{\bar{n}}) - 4\,\bar{Z}_3^{\bar{n}}\,Z_1^{\bar{n}}) = 0\}$ then

either $f(X) = h_1(X) + X^{|\bar{n}|+2}\,W_1(X)$

or $f(X) = h_2(X) + X^{|\bar{n}|+2}\,W_2(X)$

(W_1 and W_2 belong to $\mathbb{C}[[X]]$ and $W_1(0) \neq 0$ or $W_2(0) \neq 0$).

<u>Remark</u> : If $Z_1^{\bar{n}}(0) \neq 0$ the problem is very difficult ([1]) .

Next we can look for *topological convergence* results : first of all it is fundamental to choose a strategy.

<u>Hypothesis</u> : Let us suppose that there is a sequence of

$\bar{n}_i = (m, q_i, r_i) \quad (i \in \mathbb{N})$

such that :

•) $m \geq n_1 + n_2$ (number of zeros of q in $0(0, {}_o))$

•) we have a constant $M > 0$ (independant of i) such that

$$\underset{z \in \partial\, 0(0,\rho_0)}{\text{Max}} \{|h_1^{\bar{n}_i}(z)|, |h_2^{\bar{n}_i}(z)|\} \leq M \qquad \forall\, i \, .$$

If f is of the form

$$f(X) = \frac{A(X)}{\sqrt{\prod_1^t (X-x_i)}} + \frac{P(X)}{Q(X)} + \phi(X)$$

($X_i \in \mathcal{D}0(0,\rho_0)$; $P,Q \in \mathbb{C}[X]$; $Q(X) = 0$ involves X belong to $\mathcal{D}0(0,\rho_0)$; ϕ is holomorphic in $\mathcal{D}0(0,\rho_0)$).

f belongs to the class we have defined at the begining of this paragraph.

Then for every root α of q there exists a subsequence of the sequences of the roots of $\{Z_1^{\bar{n}_i}\}_i$ which converges toward α . We can locate every singularity of f in $\mathcal{D}0(0,\rho_0)$ and a direct study of the discriminant of the

$Z_1^{\bar{n}_i} y^2 + Z_2^{\bar{n}_i} y + Z_3^{\bar{n}_i} = 0$ can give us information about the nature of the singularity which we talk about (see [1],[2] for the proof and for numerical experiments).

Bibliography

1. Della Dora, J.: Contribution a l'approximation de fonctions de la variable complexe au sens de Hermite-Padé et de Hardy, Thèse présentée à l'USMG (20 juin 1980)
2. Della Dora, J., Di Crescenzo, C.: "Approximation de Padé-Hermite", in *Padé Approximation and Its Applications*. Proceedings, Antwerp, 1979, ed by L. Wuytack, Lecture Notes in Mathematics, Vol. 765 (Springer Berlin, Heidelberg, New York 1979)
3. Della Dora, J., Di Crescenzo, C.: Approximants de Padé-Hermite. Part I: Theorie, Part II: Algorithms and Applications, Submitted to Numerische Mathematik
4. Brezinski, C.: Accéleration de la Convergence en Analyse Numérique, Lecture Notes in Mathematics, Vol. 584 (Springer Berlin, Heidelberg, New York 1977)
5. Baker, G.A.: Essentials of Pade Approximants. Academic Press, New-York (1975)
6. Shafer, R.E.: On quadratic approximation. SIAM J. Numer. Anal. Vol. 11, Nr. 2 (April 1974) 447-460.
7. Gragg, W.B.: The Pade table and its relation to certain algorithms of numerical analysis. SIAM Rev. (1972) Nr. 14; 1-62.
8. Cilewicz, J.: Approximants de Padé, Lecture Notes in Mathematics, Vol. 667 (Springer Berlin, Heidelberg, New York 1978)

Some Comments About the Numerical Utilization of Factorial Series

J.P. Ramis

Université de Strasbourg, I.R.M.A., 7, Rue René Descartes
F-67084 Strasbourg Cédex, France

J. Thomann

C.N.R.S., Centre de Calcul, 23, Rue du Loess, BP 20/CR
F-67037 Strasbourg Cédex, France

The "k-summable series" theory, recently developed by one of the authors (Ramis [6], [7]), and its applications to the theory of linear differential equations and linear difference equations prove that the evaluation of "actual-problems" related functions, known by its formal asymptotic expansions, is very often reduced (when k is rational) to the evaluation of a sum of a generalized factorial series. Usually for a given problem, one can choose among an infinite family of factorial series indexed by a continuous parameter $\omega \in]\omega_0,+\infty[$ (for $0 \leq \omega < \omega_0$ the factorial series is divergent).

An investigation on the convergence of the modulus series indicates that the greater ω_0 is, the faster the series converges (cf. NÖRLUND [5], Theorem IX) ; however the "speed" of convergence remains insignificant. Nevertheless NÖRLUND points out [5](p. 378) that the convergence can be speedy if a "good choice" of ω is made. Experiments have suggested to us that, for a function originated from a differential equation, an optimal speed can be reached for $\omega > \omega_0$, $\omega - \omega_0$ being "small" ; this apparently paradoxal phenomenon seems to be related to "least term summation" properties of factorial series originated from differential equations still to be theoretically investigated. ω_0 can be theoretically determined (at least in the case of generic differential equations) but an improvement to estimate the optimal ω is still to be worked out. Here follows a summary of the theory in view of numerical investigation (k-summable series and applications).

To illustrate this, we have worked on the Euler series and the Bessel functions of integer order. Of course, some classical methods would produce more precise results than the factorial series method ; however, the latter covers a much wider field and can for instance be extended to perturbations of Euler or Bessel differential equations.

I. k-SUMMABLE SERIES AND APPLICATIONS

Introduction

The typical example of this phenomenon is given by the classical Euler equation :

$$x^2 \frac{df}{dx} + f = x .$$

If we seek after a formal solution $\hat{f} \in \mathbb{C}[[x]]$ $(\hat{f}(x) = \sum_{n\geq 0} a_n x^n)$ of this equation, we find the unique solution :

$$\hat{f}(x) = \sum_{n\geq 0} (-1)^n\, n!\, x^{n+1} .$$

On the other hand, the "variation of parameters" method gives the general solution

$$f(x) = e^{1/x}\left(\int_0^x \frac{e^{-1/t}}{t}\, dt + c\right) .$$

With $c = 0$, $f(x) = e^{1/x} \int_0^x \frac{e^{-1/t}}{t}\, dt = \int_0^{+\infty} \frac{e^{-t/x}}{1+t}\, dt$.

We can immediatly check that $\hat{f}(x)$ is the asymptotic expansion at 0 $(x>0)$ of $f(x)$. The function $\int_0^{+\infty} \frac{e^{-t/x}}{1+t}\, dt$ can be continued as an analytic function on the logarithm Riemann surface, while the asymptotic expansion remains meaningful in the sector : $|\arg x| < \frac{3\pi}{2}$. As a reminder let us call back (cf. f.ex. WASOW [9]) the :

<u>DEFINITION 1.0.1: $\overline{W}$ is a closed sector, f continuous on $\overline{W}$ and holomorphic on</u> $\overset{\circ}{W}$. f <u>uniformly admits the asymptotic expansion</u>

$$\hat{f}(x) = \sum_{n\geq 0} a_n x^n \quad \text{on } \overline{W} ,$$

<u>if, for every</u> $n \in \mathbb{N}$, $M_n > 0$ <u>exists such as</u> :

$$|x^{-n}| \left| f(x) - \sum_{p=0,\dots,n-1} a_p x^p \right| < M_n .$$

If V is an open sector, and f an holomorphic function on V , then f is said to admit an asymptotic expansion $\hat{f}$ on V , if it does so uniformly on every $\overline{W}$, W being a strict subsector of V .

$\mathcal{G}(V)$ is the $\mathbb{C}$-vector space of functions admitting an asymptotic expansion on V ,

$$\mathcal{G}(V) \xrightarrow{J} \mathbb{C}[[x]]$$
$$f \longrightarrow \hat{f} .$$

$\mathcal{G}_0(V)$ is the kernel of J . This kernel is not reduced to zero and f is not uniquely defined by its asymptotic expansion.

This fundamental defect of Poincaré's classical theory can be, as we will see, overcome in many cases by the theory of k-summable series. In more general situations, with help of the theory of Gevrey's asymptotic expansion, $\mathcal{G}_0$ shrinks to a space of functions with exponential decay (with fixed order).

The k-summable series are able to "sum" $\hat{f}$ in the following cases :

a) Solutions of linear analytic differential equations at an irregular singular point.

b) Solutions of "regular" difference equations with coefficients in $\mathbb{C}(x)$ ($k=1$ is sufficient here).

c) Problems of iteration and conjugation in the group of local analytic automorphisms tangent to the identity at the origin of $\mathbb{C}$.

The Gevrey's asymptotic expansions works for :

d) Solutions of non linear differential equation of the form :

$$G(x, f, \dots, f^{(n)}) = 0 \text{ , where } G \in \mathbb{C}\{x\}[y_0, \dots, y_n] \text{ .}$$

1. Gevrey's asymptotic expansions

The theory of "Gevrey's asymptotic expansions" has been essentially worked out by WATSON [10] , [11] , [12] and NEVANLINNA [4]. This theory is soft and as "good" as the classical theory of Poincaré on "small sectors". (The fundamental results (cf. f. ex. WASOW [9]) still remain). But on the other hand, this theory becomes "stiff" on the "big sectors", entailing summation results.

We have previously conveyed the exact sequence (V being an open sector issued from the origin of $\mathbb{C}$)(Borel-Ritt ; cf. WASOW [9]) :

$$0 \longrightarrow \mathcal{A}_0(V) \longrightarrow \mathcal{A}(V) \xrightarrow{J} \mathbb{C}[[x]] \longrightarrow 0 \text{ .}$$

Given $k>0$ and $s = 1+\frac{1}{k}$, we define : $\mathbb{C}[[x]]_s$ (formal Gevrey series of order s), $\mathcal{A}_s(V)$ (Gevrey asymptotic expansions of order s) and $\mathcal{A}_{0,s}(V) = \mathcal{A}_s(V) \cap \mathcal{A}_0(V)$.

We get the following results ($\mathrm{aper}(V)$ points to the aperture (which can be greater than 2π) of V) :

- If $\mathrm{aper}(V) < \frac{\pi}{k}$ ("small sector") :

$$\mathcal{A}_s(V) \xrightarrow{J} \mathbb{C}[[x]]_s \longrightarrow 0$$

("soft" state not quasi-analytic : cf. CARLEMAN [3]).

- If $\mathrm{aper}(V) > \frac{\pi}{k}$ ("big sector") :

$$0 \longrightarrow \mathcal{A}_0(V) \xrightarrow{J} \mathbb{C}[[x]]_s$$

("stiff" state quasi-analytic : cf. CARLEMAN [3]).

In the second case, f is uniquely determined by $J(f) = \hat{f}$ and we have a situation of "summation". This abstract summation can be achieved by integral formulas and, if k is rationnal, by a "combinatorial" method.

Définition 1.1: If $s = 1+\frac{1}{k}$ $(k>0)$ and $\hat{f}(x) = \sum_{n\geq 0} a_n x^n \in \mathbb{C}[[x]]$, $\hat{f}$ is said to verify a Gevrey condition of order s if there exist $c>0$ and $A>0$ such as :

$$|a_n| < c(n!)^{s-1} A^n .$$

We write $\hat{f} \in \mathbb{C}[[X]]\ s, A$). $\hat{f}$ is called of order $-k$ and of type $\tau = -\frac{1}{kA^k}$. (Continuation of the classical definitions of entire functions.) We can verify that $\mathbb{C}[[X]]_s$ is a $\mathbb{C}$-algebra and is stable by differentiation and composition.

Definition 1.1.2: If $s = 1+\frac{1}{k}$ $(k>0)$ and V is an open sector, $f \in \mathcal{G}(V)$ is said to admit an asymptotic expansion with a Gevrey condition of order s, for every $W<V$ (strict subsector) if there exist $C_W>0$ and $A_W>0$ such as :

$$\sup_{x\in W} |f^n(x)| \leq C_W(n!)^{s-1} A_W^n .$$

Then we write : $f \in \mathcal{G}_s(V)$ and $\mathcal{G}_{0,s}(V) = \mathcal{G}_s(V) \cap \mathcal{G}_0(V)$.

To $\hat{f}(x) = \sum_{n\geq 0} a_n x^n$, we connect $\hat{\varphi}(t) = \sum_{n\geq 0} \frac{a_n}{\Gamma(1+\frac{n}{k})} t^n$

(Formal inverse Leroy transformation).

THEOREM 1.1.3. Let $k>0$ $(s=1+\frac{1}{k})$ and V be an open sector. If $f \in \mathcal{G}(V)$ $(\hat{f} = J(f))$, the following conditions are equivalent :

(i) for every $W<V$, there exist $C_W>0$ and $A_W>0$ such that :

$$\sup_{x\in W} |f^{(n)}(x)| \leq C_W(n!)^{1/k} A_W^n \quad \text{(i.e. } f \in \mathcal{G}_s(V)) .$$

(ii) For every $W<V$, there exist $k_W>0$ and $B_W>0$ such that :

$$\sup_{x\in W} |x^{-n}|\ |f(x) - \sum_{p=0,\dots,n-1} a_p x^p| \leq K_W(n!)^{1/k} B_W^n .$$

If $\operatorname{aper}(V) < \frac{\pi}{k}$ and if $\mathbb{R}^+$ bisects V, these conditions are moreover equivalent to the two following :

(iii) $\hat{\varphi} \in \mathbb{C}\{t\}$ with a convergence radius $R>0$ and for $0<r<R$, we have :

$$f(x) - \frac{k}{x^k}\int_0^r \varphi(t)\, e^{-t^k/x^k}\, t^{k-1}\, dt \in \mathcal{C}_{0,s}(V)\ .$$

(iv) $\hat{\varphi} \in \mathbb{C}\{t\}$ and for every $W<V$, there exist :

$L>0$, $K'_W>0$, $C'_W>0$ such that :

$$\Big|f(x) - \sum_{p<L_W|x|^{-k}} a_p x^p\Big| \le C'_W\, e^{-K'_W/|x|^k} \quad \text{for every } x \in W\ .$$

(This is a "least term summation" condition).

The space $\mathcal{C}_{0,s}(V)$ is characterized by exponential decreasing of order $k = \frac{1}{s-1}$ properties :

PROPOSITION 1.1.4: Let V be an open sector of aperture $<\frac{\pi}{k}$. For $f \in \mathcal{O}(V)$ the following conditions are equivalent :

(i) $f \in \mathcal{C}_{0,s}(V)$;

(ii) $|f(x)| < \exp(\tau|x|^{-k})$ $(\tau<0)$.

2. k-summable series

Let $k>0$.

Definition 1.2.1: Let $\alpha \in \mathbb{R}/2\pi\mathbb{Z}$ and $\hat{f} \in \mathbb{C}[[x]]$.
$\hat{f}$ is called strongly k-summable in the direction α, if there exist an open sector V with vertex at 0, with bissector α and aperture $\beta>\frac{\pi}{k}$ and a function $f_\alpha \in \mathcal{C}_s(V)$ such as $J(f_\alpha) = \hat{f}$.

In this case f_α is unique and $\hat{f} \in \mathbb{C}[[x]]_s$. We can immediately see that $\hat{f}$ is also strongly k-summable in every direction β near α and that the corresponding f_β glue together. (f_α is called sum of f in the direction α .) $\hat{f}$ is called k-summable if it is so in all directions, but at most a finite number $\alpha_1, \ldots, \alpha_\ell$ ($\{\alpha_1, \ldots, \alpha_\ell\}$ will be called the singular support of $\hat{f}$, written $\Sigma(\hat{f})$.) The sums f_α glue together for $\alpha \in]\alpha_i - \frac{\pi}{2k}, \alpha_{i+1} + \frac{\pi}{2k}[$ (the radius of V_α shrinks to zero if α tends to α_i or α_{i+1}) and give a sum on a eye-shaped domain. On the intersection of such two consecutive domains (with aperture $\frac{\pi}{k}$) we get two determinations.

We can verify immediately that $\Sigma(\hat{f}) = \emptyset$ if and only if $\hat{f}$ is convergent.

We write $\mathbb{C}\{x\}_s$: the space of k-summable series $(s = 1 + \frac{1}{k})$;

$\mathbb{C}\{x\}_s$ is a $\mathbb{C}$-algebra stable for differentiation ; $\mathbb{C}\{x\}_s \subset \mathbb{C}[[x]]_s$.

PROPOSITION 1.2.2: *Let* $\hat{f} \in \mathbb{C}\{x\}_s$ *and* $\Sigma(\hat{f}) = \{\alpha_1, \ldots, \alpha_\ell\}$. *There exists a unique decomposition (modulo adding up elements of* $\mathbb{C}\{x\}$) :

$$\hat{f} = \hat{f}_0 + \hat{f}_1 + \ldots + \hat{f}_\ell \text{ , where } \hat{f}_0 \in \mathbb{C}\{x\} \text{ and } \Sigma(\hat{f}_j) = \{\alpha_j\} .$$

Let $\hat{f} \in \mathbb{C}\{x\}_s$. We suppose (if not, we first perform a rotation) that $0 \notin \Sigma(\hat{f})$.

Let $\hat{\varphi} = \Sigma \dfrac{a_n}{\Gamma(1 + \frac{n}{k})} t^n$ be the order k inverse formal Leroy transformation of $\hat{f}$. Its sum can be continued to $[0, +\infty[$ and we obtain

$f(x) = \dfrac{k}{x^k} \displaystyle\int_0^{+\infty} \varphi(t) \exp(-\dfrac{t^k}{x^k}) t^{k-1} dt$ which gives the sum of $\hat{f}$ in the direction 0 . Moreover the sum φ of $\hat{\varphi}$ can be continued on the complex plane with cuts $[a_i, \infty[$, where a_i lies in the direction α_i and φ is at most exponentially increasing of order k in the direction $\alpha \notin \Sigma(\hat{f})$.

3. Application to differential equations :

Consider the system :

$$(\Delta) \qquad \Delta(Y) = \frac{dY}{dx} - A(x)Y = 0 \; ; \text{ where } Y \in \mathbb{C}^m \text{ and } A(x) \text{ is a } (m \times m)$$

matrix whose coefficients are meromorphic. Such system admits a fundamental matrix of the form $\hat{H}(t)\, t^{\Lambda} \exp(Q(\frac{1}{t}))$, where $t^q = x$ ($q \in \mathbb{N}$; $q \leq \text{l.c.m.}(1, 2, \ldots, m) \leq m!$), Λ is a constant matrix, $Q(\frac{1}{t})$ is a diagonal matrix compounded of $\frac{1}{t}$ polynomials, $\hat{H}(t)$ is a matrix with coefficients in $\mathbb{C}[[t]]$. We write :

$$Q = \begin{pmatrix} q_1 & & & \\ & \ddots & & 0 \\ & & q_\ell & \\ 0 & & & q_{\ell+1} \end{pmatrix} ,$$

where $q_1, \ldots, q_\ell \not\equiv 0$, and $q_{\ell+1} \equiv 0$, or $Q = \begin{pmatrix} q_1 & & 0 \\ & \ddots & \\ 0 & & q_\ell \end{pmatrix}$, where $q_1, \ldots, q_\ell \not\equiv 0$, according to the different cases. We put :

k_{ij} = degree of q_{ij} $(j = 1, \ldots, \ell)$ and

$$s_{ij} = 1 + \frac{1}{k_{ij}} \; ; \; q_{ij} = q_j - q_i \; .$$

<u>THEOREM 1.3.1. There exists a decomposition</u> :

$$\hat{H} = \prod_{i,j} \hat{H}_{i,j} ,$$

<u>where the coefficients of</u> $\hat{H}_{i,j}$ <u>are in</u> $\mathbb{C}\{t\}_{s_{i,j}}$. <u>For</u> α <u>"generic"</u> ($\alpha \notin \Sigma(\hat{H}_{i,j})$) , $\hat{H}_{i,j}$ <u>has a sum</u> $H_{i\,j,\alpha}$ <u>in the direction</u> α <u>and</u> $H_\alpha = \prod_{i,j} H_{i,j,\alpha}$ <u>is a fundamental matrix for the equation</u> (Δ) .

<u>Remark</u>. In the "generic case", the q_i have all the same degree k and their terms of degree k have distinct coefficients. Then $d^\circ q_{ij} = k$, $\hat{H}$ is k-summable and we can obtain a good estimate of ω_0 (cf. introduction).

4. Factorial series

Factorial series yield a "formal" actual method to sum the k-summable series in a direction for $k \in \mathbb{Q}$. First, we reduce to $k \in \mathbb{N}$, then to $k = 1$. We make then use of :

$$\frac{1}{z^p} = \sum_{n \geq p-1} \frac{|S_n^{p-1}|}{z(z+1)\dots(z+n)} \qquad (S_n^{p-1} : \text{Stirling number of the first kind}).$$

If $\hat{f}(x)$ is k-summable in the direction α (angle θ with R^+), we transform formally $(xz = 1)$:

$$\frac{1}{z}\hat{f}\left(\frac{1}{z}\right) = \sum_{n \geq 1} \frac{a_n}{z^n} = \sum_{n \geq 0} \frac{b_{n+1}\, \omega^n\, n!}{ze^{-i\theta}(ze^{-i\theta}+\omega)\dots(ze^{-i\theta}+n\omega)} ,$$

where $b_{n+1} = \frac{1}{n!} \sum_{m=1}^{n+1} \frac{a_m}{e^{im\theta}\,\omega^{m-1}} |S_n^{m-1}|$.

PROPOSITION 1.4.1:

(i) $\hat{f}$ is 1-summable in the direction α .

(ii) $\sum_{n \geq 0} \frac{b_{n+1}\, \omega^n\, n!}{ze^{-i\theta}+\omega)\dots(ze^{-i\theta}+n\omega)}$ converges for $\omega \geq \omega_0 > 0$ in the half-plane $\mathrm{Re}(ze^{-i\theta}) > 0$.

ω_0 , which is dependent on θ , can be computed with help of the type of $\hat{f}$ (with more precision from the singularities of φ).

II. NUMERICAL APPLICATIONS

Introduction

According to I.4, when we put : $u_1 = \frac{a_1}{z}$, $u_2 = \frac{a_2}{z^2}$, ... , $u_m = \frac{a_m}{z^m}$, we can write the factorial series :

$$\frac{1}{z}\hat{f}\left(\frac{1}{z}\right) = \sum_{n \geq 0} V_{n+1} = \sum_{n \geq 0} \frac{\sum_{m=1}^{n+1} u_m \left(\frac{ze^{-i\theta}}{\omega}\right)^m |S_n^{m-1}|}{\frac{ze^{-i\theta}}{\omega}\left(\frac{ze^{-i\theta}}{\omega}+1\right)\dots\left(\frac{ze^{-i\theta}}{\omega}+n\right)} .$$

The steeply increasing of the S_n^{m-1} (Stirling numbers of the first kind) and the alternate sign of the a_m lead to a badly conditioned computation.

1. An algorithm to compute the factorial series

To avoid this, we can use the recurrent formulas of S_n^{n-1} , and we can build up a table of the form :

$$
\begin{array}{llll}
v_1^{(1)} = u_1 & v_2^{(1)} = \dfrac{u_2\, y}{y+1} & v_3^{(1)} = \dfrac{v_2^{(1)} + y v_2^{(2)}}{y+2} & v_4^{(1)} = \dfrac{2v_3^{(1)} + y v_3^{(2)}}{y+3} \\
v_1^{(2)} = u_2 & v_2^{(2)} = \dfrac{u_3\, y}{y+1} & v_3^{(2)} = \dfrac{v_2^{(2)} + y v_2^{(3)}}{y+2} & \vdots \\
v_1^{(3)} = u_3 & v_2^{(3)} = \dfrac{u_4\, y}{y+1} & \vdots & \\
\vdots & \vdots & &
\end{array}
$$

where the general term is : $v_{k+1}^{(j)} = \dfrac{(k-1)v_k^{(j)} + y v_k^{(j+1)}}{y+k}$, and the diagonal terms : $v_1^{(1)}, v_2^{(1)}, v_3^{(1)}, v_4^{(1)} \ldots v_n^{(1)}$ are the terms $v_1, v_2, \ldots, v_n$ of the factorial series.

2. Example of the Euler series

The series : $1 - 1!.x + 2!x^2 - 3!x^3 + \ldots$, introduced in I.0, after the transformation $z = \dfrac{1}{x}$, can be computed by the previous algorithm.

When $\theta = 0$, ω_0 can be computed from the singular point of the inverse Leroy's transform $\varphi(t) = \sum_{n\geq 0} (-1)^n t^n = \dfrac{1}{1+t}$ $\left(f(x) = \dfrac{1}{x}\int_0^\infty \varphi(t)\, e^{-t/x}\, dt\right)$.

When $\theta \neq 0$, ω_0 can also be computed if we first perform a rotation of angle θ (cf. Ramis et Thomann [8]).

Some results (ω_0 = Log 2 is the optimal value of ω) :

x arg x = 0	ω $\omega_0 = \log 2 = 0.693$	Value in the exponential integral table	Partial sum of the factorial series. [In the partial sum of the asymptotic series]	Pade approximant
0.1	log 2	0.915633339	0.915633339 (14 terms) [0.915276331]	0.915633339 (for[14/14])
0.5	0.85	0.7226568	0.7226592 (14 terms) $[-0.66 \times 10^6]$	
1	0.85	0.5963469	0.5963320 (74 terms) $[-0.044 \times 10^{106}]$	0.5963789 (for[20/20])

3. Example of Bessel functions of integer order ν

Asymptotic expansions at infinity :

$$\left.\begin{aligned} J_\nu(z) &= \sqrt{\frac{2}{\pi z}}\,\{P(\nu, z)\cos X - Q(\nu, z)\sin X\} \\ Y_\nu(z) &= \sqrt{\frac{2}{\pi z}}\,\{P(\nu, z)\sin X + Q(\nu, z)\cos X\} \end{aligned}\right\} \text{ when } |z| \to \infty \text{ and } |\arg z| < \pi,$$

where :

$$P(\nu, z) = 1 - \frac{(\mu-1)(\mu-9)}{2!\,(8z)^2} + \frac{(\mu-1)(\mu-9)(\mu-25)(\mu-49)}{4!\,(8z)^4} + \ldots$$

$$Q(\nu, z) = \frac{\mu-1}{8z} - \frac{(\mu-1)(\mu-9)(\mu-25)}{3!\,(8z)^3} + \ldots$$

$$X = z - (\tfrac{1}{2}\nu + \tfrac{1}{4})\pi \quad , \qquad \mu = 4\nu^2 .$$

If we compute the partial sums of the factorial series $(\omega_0 = \frac{\pi}{6})$:

ν	z	ω	Value in the tables	Partial sum of the asymptotic expansion.	Partial sum of the factorial series.
0	0.5	$\frac{\pi}{6}$	0.9384698072	0.8849 (4 terms)	0.93859334 (5 terms)
0	1	$\frac{\pi}{6}$	0.7651976866	0.7460 (3 terms)	0.76523374 (11 terms)
0	5	$\frac{\pi}{6}$	-0.1775967713	-0.1775959120 (17 terms)	-0.1775967713 (17 terms)
1	5	$\frac{\pi}{6}$	-0.3275791376	-0.3275790602 (20 terms)	-0.3275791371 (22 terms)

4. Remarks on the behaviour of the terms of the factorial series in function of θ and ω

In figures 1 and 2, the subdivisions on the x-axis correspond to the numbers of the terms in the factorial series (the number 0 is at the 6th subdivision) ; the subdivisions on the y-axis correspond to the powers of 10 and measure the modulus of each term.

We represent 50 terms of the factorial series computed from the Euler series with $|z| = 1$: in figure 1 with $\theta = 0$, in figure 2 with $\theta = \frac{\pi}{2}$ and $\arg z = \frac{\pi}{2}$.

For each value of ω , we have computed the corresponding partial sum with 50 terms of the factorial series. If $\theta = 0$, $\omega_0 = \log 2 \sim 0.693$ and if $\theta = \frac{\pi}{2}$, $\omega_0 = \frac{\pi}{3} \sim 1.047$. We can observe that : The optimal convergence is obtained for $\omega \sim 0.85$ (Fig. 1) and $\omega \sim 1.5$, i.e. for $\omega > \omega_0$, $\omega - \omega_0$ being "small". The convergence is also optimal in the opposite direction of the singular support $(\theta = 0)$.

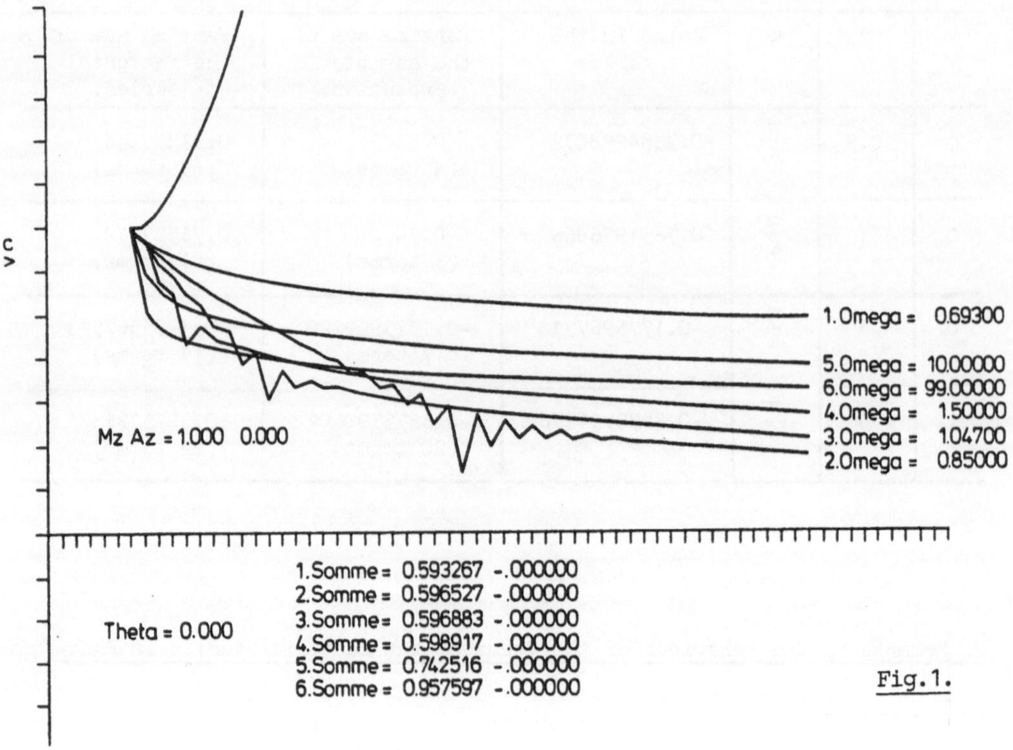
vc
1.Omega = 0.69300
5.Omega = 10.00000
6.Omega = 99.00000
4.Omega = 1.50000
3.Omega = 1.04700
2.Omega = 0.85000
Mz Az = 1.000 0.000
1.Somme = 0.593267 -.000000
2.Somme = 0.596527 -.000000
3.Somme = 0.596883 -.000000
4.Somme = 0.598917 -.000000
5.Somme = 0.742516 -.000000
6.Somme = 0.957597 -.000000
Theta = 0.000

Fig.1.

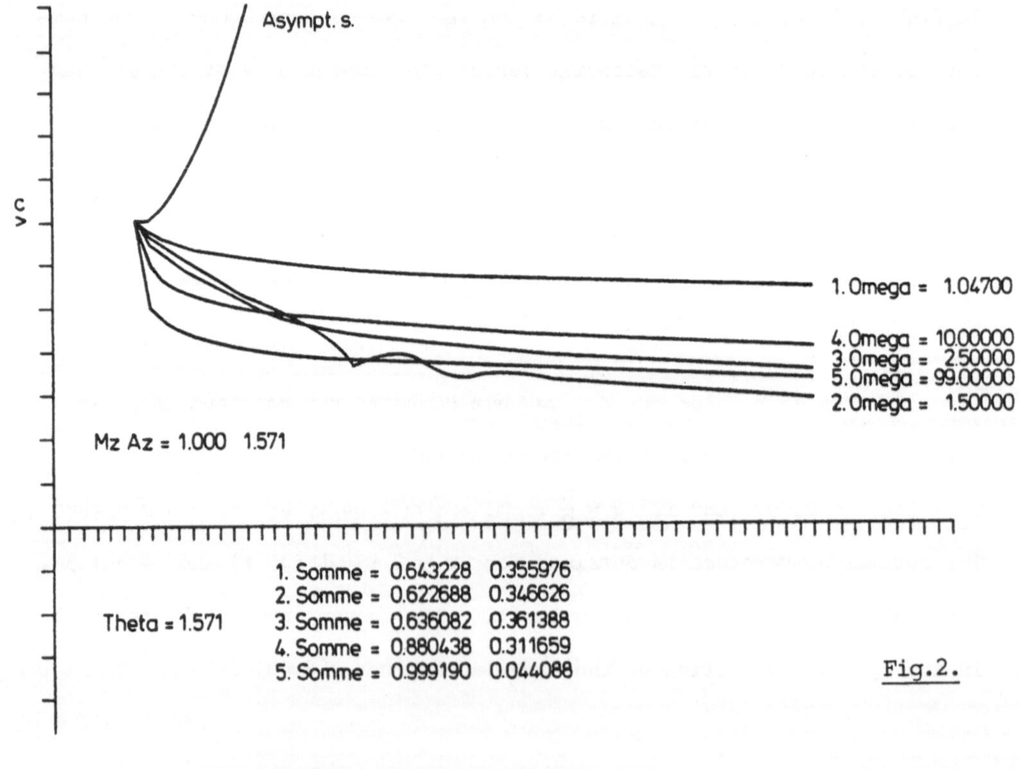
Asympt. s.
vc
1. Omega = 1.04700
4. Omega = 10.00000
3. Omega = 2.50000
5. Omega = 99.00000
2. Omega = 1.50000
Mz Az = 1.000 1.571
1. Somme = 0.643228 0.355976
2. Somme = 0.622688 0.346626
3. Somme = 0.636082 0.361388
4. Somme = 0.880438 0.311659
5. Somme = 0.999190 0.044088
Theta = 1.571

Fig.2.

Near ω_0 the curves which represents the terms of the series are broken (phenomenon connected with sign alternating). When ω grows, the curves become smoother but the convergence is less sharp.

BIBLIOGRAPHY

[1] BOREL E. Mémoire sur les séries divergentes. Annales Sc. de l'E.N.S., 3ème Série, t. 16, 1899 (9-136).

[2] BOREL E : Leçons sur les Séries Divergentes. Deuxième édition, 1928. (Gauthier-Villars, Paris.)

[3] CARLEMAN T. : Les Fonctions quasi-analytiques. (Gauthier-Villars, Paris 1926.)

[4] NEVANLINNA F. : Zur Theorie der Asymptotischer Potenzreihen. Annalen Academiae Scientiarum Fennicae, ser. A, Fasc. XII, Helsinki 1919.

[5] NÖRLUND N.E. : Sur les séries de facultés. Acta Math. 37 (1914), p. 327-387.

[6] RAMIS J.P. : Les séries k-sommables et leurs Applications. Lecture Notes in Physics 126 (Springer Verlag 1980).

[7] RAMIS J.P. : Développements asymptotiques Gevrey, séries k-sommables et applications aux équations différentielles et aux différences. A paraître (1980).

[8] RAMIS J.P., THOMANN J. : Remarques sur l'utilisation numérique des séries de factorielles. (Séminaire IMAG, Grenoble, 1980.)

[9] WASOW W. : Asymptotic Expansions for ordinary differential Equations (Interscience Publishers 1965).

[10] WATSON G.N. : A theory of Asymptotic series. Philisophical Transactions of the Royal Society of London (Ser. A), vol. CCXI (1911), p. 279-313.

[11] WATSON G.N. : The Charasteristics of Asymptotic Series. Quarterly Journal of Mathematics, vol. XLIII (1911-1912), p. 65-77 .

[12] WATSON G.N. : The transformation of an asymptotic series into a convergent series of inverse factorials. Cir. Mat. Palermo, Rend., 34 (1912), p. 41-88.

Introduction to Real Quasianalytic Classes and Continuation Problems

T. Reboul

Université de Grenoble I, Laboratoire d'Informatique et de Mathématiques Appliquées, BP : 53 X, F-38041 Grenoble Cédex, France

In this paper we point out some similar properties between analytic functions and elements of quasianalytic classes and give some examples. It seems to us that quasi-analytic classes may be a means to give a sense to the continuation of an analytic function on a path through its natural boundary.

1. Real Analytic Functions

$f : [a,b] \to \mathbf{R}$ *is a real analytic function if it is infinitely differentiable and locally coincides with its Taylor series.*

Uniqueness Property

Let f and g be defined and analytic on [a,b].

(1) if there exists a point $x_0 \in [a,b]$ such that for every natural $n, f^{(n)}(x_0) = g^{(n)}(x_0)$ then $f = g$;

(2) if $f(x) = g(x)$ in a non-void open subinterval then $f = g$.

Analytic Continuation

Let $f : [a,b] \to \mathbf{R}$, $g : [c,d] \to \mathbf{R}$, both be analytic, $I = [a,b] \cup [c,d]$, $J = [a,b] \cap [c,d]$.

(3) if there exists $x_0 \in J$ such that for every natural n, $f^{(n)}(x_0) = g^{(n)}(x_0)$ then there exists $h : I \to \mathbf{R}$, a unique analytic continuation of f and g;

(4) if $f(x) = g(x)$ on a non-void open subinterval of J then there exists $h : I \to \mathbf{R}$, a unique analytic continuation of f and g.

Characterization Theorems

(5) *PRINGSHEIM's Theorem:*

$f : [a,b] \to \mathbf{R}$ is analytic if and only if f is infinitely differentiable and there exists $k > 0$ such that for every natural n and every $x \in [a,b]$, $|f^{(n)}(x)| \leq k^n\, n!$.

(6) *BERNSTEIN's Theorem:*

Let $f : [a,b] \to \mathbf{R}$ be continuous; we note $\|f\| = \|f\|_{a,b} = \max_{x \in [a,b]} |f(x)|$,

$E_n(f) = E_n(f)_{[a,b]} = \min_p \|f-p\|$ where p runs through the set of polynomials of degree less than n.

Remark: There always exists one and only one p such that $E_n(f)_{[a,b]} = \|f-p\|_{[a,b]}$. We have the theorem [1]: f is analytic if and only if there exist $M > 0$ and $\rho \in]0,1[$ such that for every natural n, $E_n(f) \leq M\rho^n$.

Interpolation Problem

(7) Let $x_0 \in \mathbf{R}$, $\{c_n\}$ a real sequence, $\varepsilon > 0$, then there exists an analytic function $f : [x_0-\varepsilon, x_0+\varepsilon] \to \mathbf{R}$, such that $f^{(n)}(x_0) = c_n$ if

$$\overline{\lim_n} \left|\frac{c_n}{n!}\right|^{1/n} < \frac{1}{\varepsilon} .$$

(8) Let $u : [a,b] \to \mathbf{R}$, $c = (a+b)/2$, $\rho \in]0,1[$, $\gamma = [(b-a)/4][\rho+(1/\rho)]$. There exists an analytic function $f : [c-\gamma, c+\gamma] \to \mathbf{R}$ which is a continuation of u if and only if

$$\overline{\lim_n} |E_n(f)|^{1/n} < \rho .$$

Remark:

- in the case (7), $f(x) = \sum_{n=0}^{\infty} \frac{c_n}{n!} (x-x_0)^n$, $x \in [x_0-\varepsilon, x_0+\varepsilon]$.

- in the case (8), let P_n the polynomial of best approximation of u on [a,b] then $f(x) = \lim P_n(x)$, $x \in [c-\gamma, c+\gamma]$.

2. Introduction to Real Quasianalytic Classes

The Notion Introduced by J- HADAMARD, the DENJOY-CARLEMAN Classes

Let I be a real interval, $\{M_n\}$ be a real sequence of positive numbers. *The class* $C\{M\} = C\{M\}_I$ *is the set of functions* f *defined on* I, *infinitely differentiable such that there exists* $k > 0$ *satisfying for every natural* n, $\|f^{(n)}\|_I \leq k^n M_n$.

Inequalities (5) are a particular case. C{M} *is a quasianalytic class if every element* f *is uniquely defined by the sequence* $\{f^{(n)}(x_0)\}$, that is, the property (1).

The Notion Introduced by S. BERNSTEIN

Let $\{N_n\}$ be a strictly increasing sequence of naturals; $I = [a,b]$. *The class* $\{N\}_I$ *is the set of functions* f *such that there exist* $M > 0$ *and* $\rho \in]0,1[$ *satisfying every* $m \in \{N_n\}$, $E_m(f) \leq M\rho^m$.

Inequalities of theorem (6) are a particular case. $\{N\}_I$ *is a quasianalytic class if every element* f *is uniquely defined on an open, non-void subinterval, that is, condition* (2).

In the following I *and* J *are finite intervals with an interior point.*

3. Results on the DENJOY-CARLEMAN Classes

Characterization

Theorem of DENJOY-CARLEMAN: C{M} is a quasianalytic class if and only if $\sum_n 1/\beta_n$ diverges, where

$$\beta_n = \inf_{r \in \mathbb{N}} (M_{n+r})^{\frac{1}{n+r}} .$$

For equivalent conditions see [4,8,9].

Inclusion of a Class in a Class

Theorem of CARTAN and MANDELBROJT [6]*:*

$$C\{M\} \subseteq \{M'\} \quad \text{if and only if} \quad \overline{\lim_n} \left(\frac{M_n^f}{M_n'}\right)^{1/n} < \infty$$

where

$$M_n^f = \frac{1}{n^n} \sup_{r \geqq n} \frac{r^{2n}}{U(r)} \quad \text{and} \quad U(r) = \sup_{n \leqq r} \frac{r^{2n}}{n^n M_n} .$$

Note: There exists a process to compute $\{M_n^f\}$ from $\{M_n\}$, see [10] and [6]. We have the results:

- if $M_n = 0$, $C\{M\}_I$ is included in the set of polynomials of degree less than n.
- if $\lim_n (M_n/n)^{1/n} > 0$, then the analytic class is included in $C\{M\}_I$.

Algebraic Properties

- $C\{M\}_I$ is a real vector space.
- If $\overline{\lim_n}(M_n/n)^{1/n} > 0$ then $C\{M\}_I$ is stable for pointwise multiplication.
- If $p : I \to J$, $p(x) = ax + b$, $f \in C\{M\}_J$, then $f \circ p \in C\{M\}_I$.

Differential Properties

- Every primitive of $F \in C\{M\}_I$ belongs to $C\{M\}_I$.
- $C\{M\}_I$ is stable for differentiation if and only if $\overline{\lim_n}\left(M_{n+1}^f/M_n\right)^{1/n} < \infty$.

Continuation Property

Theorem: Let $f \in C\{M\}_I$, $g \in C\{M\}_J$; if there exists $x_0 \in I \cap J$ such that for every natural n, $f^{(n)}(x_0) = g^{(n)}(x_0)$ then there exists a unique $h \in C\{M\}_{I \cup J}$ which is a continuation of f and g.

Interpolation Problem

By the interpolation problem we mean the problem of the existence and of the determination of $f \in C\{M\}_I$ satisfying for every natural n, $f^{(n)}(x_0) = c_n$ where $x_0 \in I$ and $\{c_n\}$ are given, and the search for a numerical method of constructing f.

CARLEMAN [5] solved this problem using variational principles when $C\{M\}_I$ is stable for differentiation. De La Vallée Poussin gives computing processes when the existence of f is insured, and f is periodic, in particular, a process based on the least square method.

4. Results on the BERNSTEIN Classes

Characterization

BERNSTEIN's Theorem [2]: every class $\{N\}_I$ is quasianalytic.

Inclusion of a Class in a Class

Theorem: $\{N\}_I \subseteq \{N'\}_I$ if and only if there exists $\lambda \geq 1$ such that for every natural n there exists a natural r satisfying $\frac{1}{\lambda} N'_r \leq N_n \leq N'_r$.

We have the results:

- the analytic class $\{n\}_I$ is included in every class $\{N\}_I$.
- If $\overline{\lim_n}(N_{n+1}/N_n) < \infty$ then $\{N\}_I$ is the analytic class $\{n\}_I$.
- if $\{N'_n\}$ is a subsequence of N then $\{N\}_I \subseteq \{N'\}_I$.

Algebraic Properties

- $\{N\}_I$ is a real vector space.
- $\{N\}^*_I$ is stable for pointwise multiplication, where $\{N\}^*_I = \bigcup_{m \in \mathbb{N}} \{m \cdot N\}_I$.
- If p is a polynomial and $f \in \{N\}^*_J$ then $f \circ p \in \{N\}^*_I$ provided that $p(x) \in J$ when $x \in I$.

Differential Properties and Continuity

- If $f \in \{N\}_I$, then f is continuous;
- if $f \in \{N\}_I$, then every primitive of f belongs to $\{N\}_I$;
- every function of the $\{N\}_I$ is infinitely differentiable if and only if $\overline{\lim_n} \log(N_{n+1})/N_n = 0$;
- if $\overline{\lim_n} \log(N_{n+1})/N_n > 0$, then there exists $f \in \{N\}_I$ which is not infinitely differentiable.

Continuation Property

A Theorem of BERNSTEIN [2]: Let $f \in \{N\}_I$ and $g \in \{N\}_J$; if $f(x) = g(x)$ on an open, non-void subinterval of $I \cap J$, then there exists a unique $h \in \{N\}_{I \cup J}$ which is a continuation of f and g.

Let f be an element of $\{N\}_I$ and $x_0 \in I$ an extreme point. For $\varepsilon > 0$ we denote $J_\varepsilon = I \cup K_\varepsilon$ where $K = [x_0-\varepsilon, x_0+\varepsilon]$; we have the Theorem (Bernstein) [3].

There exists $\varepsilon > 0$ and a continuation $h \in \{N\}_{J_\varepsilon}$ of f if and only if there exists $\eta > 0$ such that the polynomials P_{N_n} of best approximation of f on I of degree less than N_n converge uniformly on K_η to a function g and there exists $\rho \in]0,1[$ and $M > 0$ such that $\|g-P_{N_n}\|_{K_\eta} < M\rho^{N_n}$.

Remark:

- if for every natural n, P_{N_n} is a polynomial in $(x-x_0)^2$, then we may choose η equal to the length of I. $P_{N_0}(x) + [P_{N_1}(x)-P_{N_0}(x)]+\ldots+[P_{N_n}(x)-P_{N_{n+1}}(x)]+\ldots$ converges uniformly to g on K_η;
- if for every natural n, P_{N_n} is a linear combination of odd powers of $(x-x_0)$ we have the same result;
- we have the theorem: there exists $\varepsilon > 0$ and a continuation $h \in \{N\}_{J_\varepsilon}$ of f if and only if there exists $\eta > 0$, $M > 0$ and $\rho \in]0,1[$ such that $A_{N_n}(x) = 0.5[P_{N_n}(x)+P_{N_n}(2x_0-x)]$ $B_{N_n}(x) = 0.5[P_{N_n}(x)-P_{N_n}(2x_0-x)]$ are uniformly convergent, respectively, to functions A and B on K_η and $\|A-A_{N_n}\|_{K_\eta} \leqq M\rho^{N_n}$, $\|B-B_{N_n}\|_{K_\eta} \leqq M\rho^{N_n}$.

5. Examples

- Let

$$a(x) = \exp\left(-\frac{1}{x}\right) \quad , \quad x \in]0,1] \quad , \quad a(0) = 0 \quad , \quad I = [0,1] \quad .$$

a is infinitely differentiable on I, analytic on]0,1] and does not belong to any quasianalytic class $C\{M\}_I$.

- Let

$$b(x) = \sum_{n=1}^{\infty} \exp\left(-\frac{2^n}{n}\right) \cos(2^n x) \quad ,$$

then b belongs to the quasianalytic class $C\{M\}_I$ where $M_n = (n \log n)^n$ and is nowhere analytic.

- Let us define $\log_1 = \log$, $\log_{p+1} = \log \circ \log_p$; then

$$c(x) = \sum_{n=k}^{\infty} \exp\left(-\frac{3^n}{\log_p(n)}\right) \cos(3^n x)$$

belongs to the quasianalytic class $C\{M^P\}_I$ where $M_n^P = [n \log_{P+1}(n)]^n$ and is nowhere analytic.

- Let us define $T_n(x) = \cos n \arccos(x)$, $x \in [-1,1] = J$, then

$$d(x) = \sum_{n=1}^{\infty} \exp\left(-\frac{2^{n^2}}{n}\right) T_{2^{n^2}}(x)$$

is infinitely differentiable on J, belongs to the class $\{2^{n^2}\}_J$ and is nowhere analytic.

- Let

$$e(x) = \sum_{n=1}^{\infty} \exp\left(-\frac{3^n}{\log n}\right) T_{3^n}(x) \quad ,$$

then e belongs to the quasianalytic class $C\{(n \log n)^n\}_J$ and does not belong to any Bernstein class $\{N\}_J$.

- Let

$$f(x) = \sum_{n=k}^{\infty} \exp\left(-\frac{3^n}{\log_P(n)}\right) T_{3^n}(x) \quad ,$$

f belongs to the quasianalytic class $C\{M^P\}_J$ and does not belong to any class $\{N\}_J$.

- Let $N_0 = 1$, $N_{n+1} = 3^{N_n}$ and let r be a natural greater than 0, then

$$g(x) = \sum_{n=1}^{\infty} 3^{-2rN_n} T_{N_{n+1}}(x)$$

belongs to the class $\{N\}_J$, is r-1 continuously differentiable but is not r continuously differentiable.

- Let

$$h(x) = \sum_{h=1}^{\infty} 3^{-(\log n)N_n} T_{N_{n+1}}(x) \quad ,$$

then h is infinitely differentiablé, belongs to $\{N\}_J$ and does not belong to any quasianalytic class $C\{M\}_J$.

We now give some proofs.

$b \in C\{M\}_I$ *and is nowhere analytic:* Let

$$u_n(r) = 2^{rn} \exp\left(-\frac{2^n}{n}\right) \quad , \quad \theta_n(r) = \log \frac{u_{n+1}(r)}{u_n(r)} \quad , \quad A(r) = \sum_{n=1}^{\infty} u_n(r) \quad ,$$

then

$$|b^{(r)}(x)| \leq A(r) \quad \text{and} \quad \theta_n(r) = r \log 2 - \frac{2^n}{n+1}\left[1 - \frac{1}{n}\right] \quad .$$

Let

$$N_r = \frac{\log r + \log_2(r)}{\log_2} \quad ;$$

for every $n \geqq N_r + 1$ we have

$$\frac{2^{n-1}}{n-1} \geqq \log_2 \frac{r \log r}{\log r + \log_2 r}$$

because $2^t/t$ increases for $t \geqq 1/\log_2$; then we have

$$\theta_n(r) = r \log 2 - 2 \frac{2^{n-1}}{n-1} \frac{n-1}{n+1} \left[1 - \frac{1}{n}\right] \leqq r \log 2 \left[1 - 2 \frac{\log r}{\log r + \log_2 r + \log 2} \left(1 - \frac{\log 2}{\log r}\right)\right] \quad ;$$

then there exists a natural $r_0 > 2/\log 2$ such that for every $r \geqq r_0$

$$\theta_n(r) < -r \frac{\log 2}{2} \quad ,$$

$$A(r) < \sum_{n \leqq N_r} 2^{rn} \exp\left(-\frac{2^n}{n}\right) < 4 \cdot N_r \cdot 2^{r \cdot N_r} = 4 \cdot \frac{\log r + \log_2(r)}{\log 2} (r \log r)^r \quad .$$

This leads to the existence of a constant k such that for every integer $r > 1$, $A(r) < k^r (r \log r)^r$, thus $b \in C\{M\}_I$.

To see that b is nowhere analytic consider

$$B(Z) = \sum_{n=1}^{\infty} \exp\left(-\frac{2^n}{n}\right) Z^{2^n} \quad .$$

This series converges on the closed unit disk $\bar{D}$, and from HADAMARD's gap theorem the unit circle S is a natural boundary for B. Now we need the

Lemma: Let F(Z) be analytic on the open unit disk D, continuous on $\bar{D}$ and such that S is a natural boundary for F, then $f : \mathbb{R} \to \mathbb{R}$ defined by $f(x) = \text{Real } F[\exp(ix)]$ is nowhere analytic.

It is sufficient to see that f is not analytic at 0. Suppose the contrary. Let

$$G(Z) = f(\log Z) \quad , \quad \log Z = -\sum_{n=1}^{\infty} \frac{(1-Z)^n}{n} \quad ;$$

there exists an open and connected neighborhood V of 1 on which G is analytic. Since $\text{Real } [F(Z) - G(Z)] = 0$, $Z \in V \cap S$, by Schwarz's principle there exists an analytic continuation H of $F(Z) - G(Z)$ in a neighborhood of 1, thus 1 is a regular point for F, contradiction. We conclude that $b(x) = \text{Real } B[\exp(ix)]$ is nowhere analytic.

$c \in C\{M^P\}_I$ *and is nowhere analytic:* We use the same arguments, but we have to replace N_r by

$$K_r = \frac{\log r + \log_{p+1}(r)}{\log 3}$$

and the condition

$$n \geqq N_r + 1 \qquad \text{by} \qquad n \geqq K_r + 1 \quad .$$

$d \in \{2^{n^2}\}_J$ *is infinitely differentiable and nowhere analytic:* We have

$$E_{2^{m^2}}(d)_J \leqq \sum_{n=m+1}^{\infty} \exp\left(-\frac{2^{n^2}}{n}\right) < \sum_{n>2^{m^2}} \exp(-n) < e \cdot e^{-2^{m^2}}$$

thus $d \in \{2^{n^2}\}_J$. The theorem of Markoff asserts that if a polynomial p is of degree n then

$$\|P'\|_J \leqq n^2 \|P\|_J \quad ;$$

then, since

$$\sum_{j=1}^{\infty} 2^{n^2 \cdot r} \exp\left(-\frac{2^{n^2}}{n}\right)$$

converges for every natural r, d is infinitely differentiable and d is nowhere analytic since d(cos x) is the real part of an analytic function with natural boundary S.

$e \in C\{M\}_J$ *and does not belong to any class* $\{N\}_J$*:* To see that $e \in C\{M\}_J$ use Markoff's theorem and the first proof. Now, for every natural $m > 0$ there exists a natural r such that $3^r \leqq m < 3^{r+1}$. Let

$$P_m = \sum_{n=1}^{r} \exp\left(-\frac{3^n}{\log n}\right) T_{3^n}(x) \quad ,$$

P_m is a polynomial of degree less than m. If

$$x = \cos\left(\frac{\pi k}{3^{(r+1)}}\right)$$

where k is a natural less than 3^{r+1}, then

$$(f-P_m)(x) = \sum_{n=r+1}^{\infty} \exp\left(-\frac{3^n}{\log n}\right) T_{3^n}(x) = (-1)^k \sum_{n=r+1}^{\infty} \exp\left(-\frac{3^n}{\log n}\right) = (-1)^k \|f-P_m\|_I \quad .$$

We use the following theorem: P_m *is the polynomial of best approximation of degree less than* m *of* f *on* J*, if and only if there exists* $x_0 < x_1 < \dots < x_{m+1}$ *elements of* J*, such that*

$$|(f-P_m)(x_i)| = E_m(f)_J \quad \textit{and} \quad (f-P_m)(x_j) = -(f-P_m)(x_{j+1})$$

for every natural $j \leq m$, to conclude that

$$E_m(e) = \sum_{n=r+1}^{\infty} \exp\left(-\frac{3^n}{\log n}\right) > \exp\left(-\frac{3^{(r+1)}}{\log(r+1)}\right)$$

and thus

$$\left(E_m(e)\right)^{1/m} > \exp\left(-\frac{3^{r+1}}{\log(r+1)}\right) \geq \exp\left(-\frac{3}{\log(r+1)}\right) \quad ;$$

the last quantity tends to 1 when m tends to $+\infty$, thus e does not belong to any class $\{N\}_J$.

$f \in C\{M^p\}_J$ *and does not belong to any class* $\{N\}_J$: The arguments are similar to the preceding ones.

$g \in \{N\}_J$ is r-1 *continuously differentiable but not* r *continuously differentiable.* We have the results:

$$T_n^{(k)}(1) = \frac{n^2(n^2-1)\ldots(n^2-(k-1)^2)}{1\cdot 3\ldots(2k-1)} \quad \text{and} \quad |T_n^{(k)}(x)| \leq T_n^{(k)}(1) \quad ,$$

for every $x \in J$. Then it is easy to prove that g is r-1 continuously differentiable. q is not r continuously differentiable because for every $h \in]0,1[$

$$\frac{T_m^{(r-1)}(1) - T_m^{(r-1)}(1-h)}{h} > 0$$

and

$$\frac{g^{(r-1)}(1) - g^{(r-1)}(1-h)}{h} = \sum_{n=1}^{\infty} 3^{-2\cdot r\cdot N_n} \cdot \frac{T_{N_n}^{(r-1)}(1) - T_{N_n}^{(r-1)}(1-h)}{h}$$

thus

$$\lim_{h\to 0^+} \frac{g^{(r-1)}(1) - g^{(r-1)}(1-h)}{h} = \sum_{n=1}^{\infty} 3^{-2\cdot r\cdot N_n} T_{N_n}^{(r)}(1) = +\infty \quad .$$

$h \in \{N\}_J$ *is infinitely differentiable and does not belong to any quasianalytic class* $C\{M\}_J$. It is easy to see that $h \in \{N\}_J$ and is infinitely differentiable. We have

$$|h^{(r)}(x)| \leq h^{(r)}(1) = \sum_{h=1}^{\infty} 3^{-(\log n)N_n} T_{N_{n+1}}^{(r)}(1) \quad .$$

Let r be the integer such that

$$e^r < n_r < e^r + 1$$

then

$$h^{(r)}(1) \geq \frac{\left(3^{-\log(e^r+1)N_{n_r}} \cdot \left(3^{2\cdot N_{n_r}} - 1\right) \cdot \left(3^{2\cdot N_{n_r}} - 2^2\right) \ldots \left(3^{2\cdot N_{n_r}} - (r-1)^2\right)\right)}{1\cdot 3\cdot 5 \ldots (2r-1)} ,$$

thus

$$h^{(r)}(1) \geq \left(3^{-(r+1)\cdot N_{n_r}} \, 3^{2\cdot r\cdot N_{n_r}}\right) \Big/ (4r)^r \quad ;$$

then let

$$a_r = |h^{(r)}(1)|^{1/r} \geq \frac{3^{[(r-1)/r]N_{n_r}}}{4r} ,$$

then we have

$$\sum_r \frac{1}{a_r} < \infty \quad ;$$

thus h does not belong to any quasianalytic class $C\{M\}_J$.

6. Comparison of Classes and Quasianalytic Continuation

A theorem of Mandelbrojt [7]: Let f be an infinitely differentiable function on the compact interval I, then there exists two quasianalytic classes $C\{M_1\}_I$, $C\{M_2\}_I$ and $f_1 \in C\{M_1\}_I$, $f_2 \in C\{M_2\}_I$ such that $f = f_1 + f_2$.

A theorem of Markushevich [12]: Let I be a compact interval and $f: I \to \mathbb{R}$ be a continuous function, then there exists two classes $\{N_1\}_I$, $\{N_2\}_I$ and $f_1 \in \{N_1\}_I$, $f_2 \in \{N_2\}_I$ such that $f = f_1 + f_2$.

From these theorems we can deduce

- There exists two quasianalytic classes $C\{M_1\}_I$, $C\{M_2\}_I$ (respectively, $\{N_1\}_I$, $\{N_2\}_I$) which cannot both be included in a quasianalytic class $C\{M\}_I$ (respectively, $\{N\}_I$).
- The solution of the interpolation problem depends on the quasianalytic class, that is, there exists two classes $C\{M\}_I$ and $C\{M'\}_I$ such that there exists $f \in C\{M\}_I$ and $g \in C\{M'\}_I$ satisfying $f \notin C\{M'\}_I$, $g \notin C\{M\}_I$ and for every natural n, $f^{(n)}(x_0) = g^{(n)}(x_0)$.
- Let f belong to the quasianalytic class $C\{M\}_I$ (respectively, $\{N\}_I$), let $C\{M'\}_J$ be a quasianalytic class and $g \in C\{M'\}_J$ (respectively, $\{N'\}_J$) be a continuation of f. We ask the question: Does g depend on $\{M_n'\}$ (respectively, $\{N_n'\}$)? The answer is yes even if $C\{M\}_I \subset C\{M'\}_I$ (respectively, $\{N\}_I \subset \{N'\}_I$).

7. A Continuation Problem

Let

$$F : \bar{D} \to \mathbb{C} \quad , \quad F(Z) = \sum_{n=1}^{\infty} \exp\left(-\frac{2^n}{n}\right) Z^{2^n} \quad ,$$

F is analytic on D, continuous on $\bar{D}$, the unit circle is a natural boundary for F and there exists $k > 0$ such that for every natural $n \neq 0$ and every $z \in \bar{D}$ we have the inequalities:

$$|F^{(n)}(z)| \leq k^n (n \log n)^n \quad .$$

We set the problems:

1) Find the segments I, $I \cap \bar{D} \neq \emptyset$, on which F may be continued into the class $C\{(n \log n)^n\}_I$.

2) Find a wider set U, which includes $\bar{D}$, on which F has a unique continuation.

3) Find an efficient algorithm to construct the continuation F on U.

References

Serge Bernstein: Leçons sur les propriétés extrêmales et la meilleure approximation des fonctions analytiques d'une variable réelle, Gauthier-Villars, Paris 1926
1 p.112-113
2 p.163
3 p.166

Torsten Carleman: Les fonctions quasi-analytiques, Gauthier-Villars, Paris 1926
4 p.61
5 p.73
6 *H. Cartan et S. Mandelbrojt:* Solution du problème d'équivalence des classes de fonctions indéfiniment dérivables. Acta Mat. T72 (1940) p.31-49

S. Mandelbrojt:
7 Sur les fonctions indéfiniment dérivables. Acta Mat. T72 (1940) p.15
8 Séries de Fourier et classes quasianalytiques de fonctions. Gauthier-Villars, Paris (1935)
Séries adhérentes régularisation des suites applications
9 p.101-103
10 p.14
11 *Charles de la Vallée Poussin:* Quatre leçons sur les fonctions quasi-analytiques de variable réelle. Bulletin de la Société Mathématiques de France T52 (1924) p.201-203
12 *A.F. Timan:* Theory of approximation of functions of a real variable. Pergamon Press 1963, p.374

1.2 Critical Phenomena in Dynamical Systems

Groups Transformations and Critical Asymptotics Applications to Non-Linear Differential and Partial Derivative Equations

J.R. Burgan, M.R. Feix, E. Fijalkow, M.P. Moraux, and A. Munier

C.R.P.E., C.N.E.T., C.N.R.S., Université d'Orléans et CEA Limeil
F-45045 Orléans Cédex, France

I. Introduction

This work is connected to different fields of physics and mathematics. A first and obvious reason is that non-linear differential and partial derivative equations are found in many - if not all - fields of physics. But, more important, as a methodology the use of transformations groups is closely connected to other methodologies which in the recent years have played an important role in theoretical physics. The first deals with the concept of self-similarity and dimensional analysis as exposed in the classical book by Sedov (1). In fact the self-similar group technique is a straightforward continuation of the dimensional analysis. A second methodology which has found important applications in statistical mechanics and field theory is the so-called renormalisation technique (2). We will deal with quite similar ideas in our force and time renormalisation concept. Finally we will see that these transformations are also very useful to optimize numerical analysis method.

This paper will introduce two kinds of group transformations, respectively called invariant and quasi invariant. Altough we will see that they complement nicely each other to treat different problems, they introduce two philosophie completely different in their spirit. The first one is based on a precise mathematical theory about the structure of solutions of equations which are strictly invariant for some group transformations. The concept of quasi invariance is based on the known properties of a kind of equations and the eventual possibility of getting rid (in a certain sense) of the difficult terms. In certain cases we will be able to clarify this concept but additionnal work is certainly needed. On the other hand the very flexible character of the quasi-invariance transformations makes it quite useful in many problems.

II. The Importance of being Invariant

We consider the following nonlinear equation

$$\frac{d^2x}{dt^2} + k\left(\operatorname{sgn}\frac{dx}{dt}\right)\left|\frac{dx}{dt}\right|^m + A(\operatorname{sgn} x)\,|x^p| = 0 \qquad |1|$$

sgn x is the function sign of x and is equal to + 1 if x > 0 and − 1 if x < 0. Let us look if the equation remains invariant under the following (rescaling) transformation

$$t = a^{\alpha}\,\bar{t} \qquad\qquad x = a^{\beta}\,\bar{x} \quad . \qquad |2|$$

We see that |1| remains invariant for all values of a provided

$$\beta - 2\alpha = m(\beta - \alpha) = p\,\beta \quad ; \qquad |3|$$

|3| implies that $p = m/(2-m)$ and that $\beta = (m-2)\,\alpha/(m-1)$. We will suppose in order to have $p > 0$ that $0 < m < 2$ and that $m \neq 1$. From this simple transformation leaving the equation invariant we deduce two kinds of results.

The first, analytical, is based on the fact that $I = x/t^{\beta/\alpha} = \bar{x}/\bar{t}^{\beta/\alpha}$ is an invariant of the transformation. Consequently one peculiar solution is

$$X = K\left(\frac{t}{T}\right)^{\frac{m-2}{m-1}} \qquad |4|$$

where K is obtained by plugging |4| in |1|.

The second result is of numerical nature. Let us call $x(x_o, v_o, t)$ the solution at time t of |1| when $x(0) = x_o$ and $\frac{dx}{dt}(t = 0) = v_o$. The existence of the transformation group when $p = m/(2-m)$ implies

$$X\left(\lambda^{\frac{m-2}{m-1}} x_o,\ \lambda^{-\frac{1}{m-1}} v_o,\ \lambda t\right) = \lambda^{\frac{m-2}{m-1}} X(x_o, v_o, t) \ , \qquad |5|$$

λ being an arbitrary real positive number $0 < \lambda < \infty$. |5| indicates that the space of initial conditions can be reduced from dimension 2 to dimension 1. More precisely suppose that we want to have an abacus of the solutions. We can take systematically $v_o = 1$ and concentrate our numerical effort on a refined sampling on x_o. When we will have to treat the initial condition v_o we will select λ such that $\lambda^{-1/(m-1)} = v_o$ which will indicate the scaling factors for time and space. (Notice, however that since $\lambda^{-1/(m-1)}$ is systematically positive we must treat two velocities $v_o = 1$ and $v_o = -1$).

Let us finally point out that |5| is nothing else but a kind of superposition principle for non-linear equation. For example we look at the form taken by |5| when $m \to 1$. We must take also λ close to 1 and we write $\lambda = 1 + \eta\varepsilon$ and $m = 1 + \varepsilon$

$$\lambda^{\frac{m-2}{m-1}} = (1 + \eta\varepsilon)^{-\frac{1}{\varepsilon}+1} \to \exp - \eta = \mu \qquad \text{when } \varepsilon \to 0$$

$$\lambda^{-\frac{1}{m-1}} = (1 + \eta\varepsilon)^{-\frac{1}{\varepsilon}} \to \exp - \eta = \mu \qquad \text{when } \varepsilon \to 0 \quad ;$$

|5| is now written

$$X(\mu x_o, \mu v_o, t) = \mu\, X(x_o, v_o, t) \qquad |6|$$

while |1| becomes the linear equation

$$\frac{d^2x}{dt^2} + k\frac{dx}{dt} + Ax = 0 \ . \qquad |7|$$

|6| is the superposition principle for the linear equation |7|.

III. Self-Similar Groups and Partial Derivative Equations

We take the example of the non-linear one dimensional heat equation

$$\frac{\partial\psi}{\partial t} = \frac{\partial}{\partial x}\left[\psi^{s}\frac{\partial\psi}{\partial}\right] \quad . \qquad |8|$$

The stretching transformations

$$t = a^{\alpha}\,\bar{t} \qquad x = a^{\beta}\,\bar{x} \qquad \psi = a^{\gamma}\,\bar{\psi}$$

leave |8| invariant provided we take $\gamma = (2\beta - \alpha)/s$, α and β being arbitrary. We get the two invariants

$$\xi = \frac{x}{(t/T)^{\beta/\alpha}} = \frac{\bar{x}}{(\bar{t}/T)^{\beta/\alpha}} \quad \text{and} \quad \frac{\psi}{(t/T)^{\gamma/\alpha}} = \frac{\bar{\psi}}{(\bar{t}/T)^{\gamma/\alpha}} = \Phi \quad . \qquad |9|$$

Writing $\beta/\alpha = \omega$ and taking the two invariants ξ and Φ respectively as new variable and new function we can state that the solutions Φ invariant under the transformation are functions only of the new variable ξ, i.e. $\Phi = \Phi(\xi)$. It is easily verified that introducing |9| in |8| we obtain a new differential equation for Φ

$$\frac{d}{d\xi}\left[\Phi^{s}\frac{d\Phi}{d\xi}\right] = \frac{1}{T}\left[\frac{2\omega - 1}{s}\,\Phi - \omega\xi\,\frac{d\Phi}{d\xi}\right] \quad . \qquad |10|$$

The "balance" of this transformation is the following

→ On the positive side : we have eliminated one variable by compacting the abscissa x and the time t in one new variable ξ.

→ On the negative side : At initial time (which in self-similar problem must be always taken at time $t = T$) the initial conditions are – of course – imposed : we can simply choose $\Phi(t = T) = \psi(t = T)$ and $\frac{d\Phi}{d\xi}(t = T) = \frac{\partial\psi}{\partial x}(t = T)$ at an arbitrary point (for example $\xi = x = 0$). The initial profile of ψ is then completely determined.

The question is now to decide if these initial conditions are close to the one we wanted to study or at least if they are interesting. Consequently we will study the stability of these initial conditions (i.e. what happens if we choose slightly different initial conditions and how the self-similar solution behaves: as an attractor or a repulsor?).

In |10| we have an arbitrary parameter ω. The solution corresponding to a critical value ω_c plays a central role. We will come back to this point in VI.

IV. The Quasi-Invariance Concept

As already said the concept is connected to the possibility of keeping a certain formalism (or sometimes a certain form) for the equation. It is best to take the following example. We consider the ODE (ordinary differential equation)

$$\frac{d^2x}{dt^2} + A\,\omega^2(t)\,x + B\,\lambda(t)\,x^3 = 0 \qquad |11|$$

where $\omega^2(t) = (1+\Omega t)^{-\mu}$, $\lambda(t) = (1+\Omega t)^{-\nu}$ μ and ν being two positive real numbers. |11| is an equation describing the Hamiltonian motion of a particle in a field of force. We introduce a generalised canonical transformation. In this respect

→ we transform the time t into a new time $\theta(t)$

→ we rescale $x(t)$ with $x(t) = \xi(\theta)\,C(t)$.

The idea is to describe the time variation of x with two terms. The first $C(t)$ must be chosen in order to simplify the equation on $\xi(\theta)$ still keeping a Hamiltonian formalism. Let us express d^2x/dt^2 as function of ξ, $d\xi/d\theta, d^2\xi/d\theta^2$ and the derivatives of θ with respect to time

$$\frac{dx^2}{dt^2} = \left[2\,\frac{dC}{dt}\,\frac{d\theta}{dt} + C\,\frac{d^2\theta}{dt^2}\right]\frac{d\xi}{d\theta} + C\,\frac{d^2\xi}{d\theta^2}\left(\frac{d\theta}{dt}\right)^2 + \xi\,\frac{d^2C}{dt^2} \ . \qquad |12|$$

To express $\frac{d^2\xi}{d\theta^2}$ as a new function of ξ and ξ^3 we must cancel the first term of the R.H.S. of |11|. This imposes the relation between θ, t and $C(t)$. More precisely we must have $d\theta = dt\,C^{-2}$. Introducing in $dx/dt = C d\xi/d\theta\; d\theta/dt + \xi\; dC/dt$ such that $d^2x/dt^2 = E$ we find the following rescaling in the new phase space ξ, η ($\eta = d\xi/d\theta$). If

$$\frac{d^2x}{dt^2} = E \ , \qquad |13.a|$$

$$x = \xi\,(\theta)\,C(t) \qquad v = \frac{1}{C}\frac{d\xi}{d\theta} + \xi\,\frac{dC}{dt} \qquad |13.b|$$

$$\eta = \frac{d\xi}{d\theta} \qquad \frac{d\eta}{d\theta} = \frac{d^2\xi}{d\theta^2} = \varepsilon \qquad |13.c|$$

$$d\theta = \frac{dt}{C^2} \qquad \varepsilon = C^3\,E - C^3\,\frac{d^2C}{dt^2}\,\xi \ . \qquad |13.d|$$

|13.d| gives the rescaling relations both for the time and for the force. Note that |13.d| introduces a "transformation field" $-\,C^3\,d^2C/dt^2\,\xi$. We are now dealing with a group transformation where each element is characterised by an arbitrary function $C(t)$ (which must not vanish in order that ξ a remains at finite distance. Applied to equation |11| transformations |13| give the "quasi-invariant" equation

$$\frac{d^2\xi}{d\theta^2} + \left(A\,\omega^2\,C^4 + C^3\,\frac{d^2C}{dt^2}\right)\xi + B\,\lambda C^6\,\xi^3 = 0 \ . \qquad |14|$$

V. Time Renormalisation and Time Compression

|14| exhibits the same kinds of difficulty as |11|. But we have at our disposal the function C(t) which is arbitrary. A first idea is to select C(t) such that the interval 0 – ∞ of t is "compacted" in a finite interval 0 – θ_1. This operation will be called renormalisation of time. A priori it is enough to take C(t) going to infinity as $t^{1/2+\varepsilon}$ where $\varepsilon > 0$. But of course we gain nothing if the time renormalisation is obtained at the expense of forces increasing without limit when $\theta \to \theta_1$. We will consequently select C(t) accordingly the following rule.

Take C(t) increasing with t as fast as possible but keeping the coefficients of ξ and ξ^3 in |14| finite.

We immediately see that taking $C(t) = (1 + \Omega t)^{1/2}$ gives for the "transformation field" $(\Omega^2/4)\xi$. Taking $C(t) = (1 + \Omega t)^{\gamma}$ with $\gamma > 1/2$ will give a transformation field going to infinity. Consequently – very often – we will have to content ourselves with a logarithmic compression of the time since $d\theta = dt/(1 + \Omega t)$ gives $\Omega\theta = \log(1 + \Omega t)$. Moreover we must check that the rescaled terms $A\omega^2(t)C^4$ and $B\lambda(t)C^6$ do not blow up. On the other hand if they decrease fast enough in order to allow taking $C(t) = 1 + \Omega t$, the transformation field vanishes and the time renormalisation is possible.

Consequently we now select the power γ of $C(t) = (1 + \Omega t)^{\gamma}$ according to the different values of μ and ν. The details can be found in (3).

The method gives the bifurcation border lines for the asymptotic solutions in the parameter phase space. Notice that we can treat |11| using the self-similar stretching groups only if $\mu = 2$.

VI. Quasi Invariance and Partial Derivative Equations

We must now show how works the quasi-invariance concept on a partial derivative equation. We are going to treat |8| again to show the connection between the self-similar and the quasi-invariance group techniques. We do not give any details, which can be found in |4|.

As in the preceeding paragraph we introduce a new time $\theta(t)$ and rescale x with $x = \xi\, C(t)$. We also rescale the function ψ with $\psi(x, t) = A(t)\, \Phi(\xi, \theta)$. We impose that in the new space time $\Phi(\xi, \theta)$ keeps the important property of heat conservation. Namely if $\int\psi(x, t)\, dx = Q$ is a finite quantity at initial time we know from |8| that Q is time invariant. We impose the same relation for $\int\phi(\xi, \theta)\, d\xi$ since

$$\int\psi(x, t)\, dx = Q = A(t)\, C(t) \int\phi(\xi, \theta)\, d\xi \quad ; \qquad |15|$$

we must have $A(t)\, C(t) = 1$. We obtain finally the transformed equation

$$\frac{\partial\phi}{\partial\theta} = \frac{\partial}{\partial\xi}(\phi^s \frac{\partial\phi}{\partial\xi}) + \frac{1}{s+2}\frac{d\,C^{s+2}}{dt}\frac{\partial}{\partial\xi}(\xi\,\phi)$$

$$\frac{d\theta}{dt} = \frac{1}{C^{s+2}} \;; \qquad |16|$$

we see that in that case we can only get a time logarithmic compression. Taking $C^{s+2} = 1 + \Omega t$ we have

$$\frac{\partial\phi}{\partial\theta} = \frac{\partial}{\partial\xi}(\phi^s \frac{\partial\phi}{\partial\xi}) + \frac{\Omega}{s+2}\frac{\partial}{\partial\xi}(\xi\,\phi)$$

$$\Omega\theta = \log(1 + \Omega t) \ . \qquad |17|$$

The connection with the self-similar equation |10| is obtained by taking the free parameter ω equal to $(s + 2)^{-1}$ and identifying Ω with T^{-1} (Notice also that we just replace t/T by $1 + \Omega t$ which is just the needed time translation since in the self-similar group technique we take the time origin at t = T). |10| indeed can now be written

$$\frac{d}{d\xi}(\phi^s \frac{d\phi}{d\xi}) + \frac{1}{(s+2)T}\frac{d}{d\xi}(\xi\,\phi) = 0 \ . \qquad |18|$$

The self-similar solution |18| appears as the steady state of the new heat diffusion equation: As pointed earlier this solution for the critical parameter $\omega = \omega_c = (s + 2)^{-1}$ not only plays a central physical role but also is the only one allowing a complete integration of |18|. Taking a symmetric solution in ξ ($d\phi/d\xi = 0$ for $\xi = 0$) we obtain

$$\phi_o(\xi) = K^{1/s}(1 - \Xi^2)^{1/s} \quad \text{for } |\Xi| < 1 \qquad \Xi = \xi/\xi_c$$

$$\phi_o(\xi) = 0 \quad \text{for } |\Xi| > 1 \qquad \xi_c^{\,2} = 2K(s + 2)\,T/s \ . \qquad |19|$$

The quasi-invariant equation |17| is now specially interesting for studying the solution $\phi_o(\xi)$ when the initial conditions are not quite those given by |19|. Especially we study the linear stability of $\phi_o(\xi)$, i.e. we take $\phi(\xi, \theta) = \phi_o(\xi) + \rho(\xi, \theta)$ supposing that $\rho \ll \phi_o$. It can be shown that taking for $\rho(\Xi, \theta)$ a Gegenbauer polynominal expansion of the form

$$\rho(\Xi, \theta) = \sum_{n=0}^{\infty} g_n(\theta)\, C_n^{\alpha}(\Xi)\,(1 - \Xi^2)^{\alpha - \frac{1}{2}} \qquad \text{with } \alpha = \frac{1}{s} - \frac{1}{2} \qquad |20|$$

in |20| C_n^{α} is the Gegenbauer polynomial of order n. We obtain for $g_n(\theta)$

$$g_n(\theta) = g_n(\theta = 0)\exp - n(n + 2\alpha)\frac{s\theta}{2(s + 2)\,T} \ . \qquad |21|$$

All the coefficients are damped except $g_o(\theta)$. Since $C_o^{\alpha}(\Xi) = 1$, its time invariance corresponds to the constancy of $\int\phi(\xi, \theta)\,d\xi$, a property formed in the quasi-invariant equation.

VII. Quasi Invariance and Schroedinger Equation

Altough the Schroedinger equation looks formally very close to the heat equation, we must remember its Hamiltonian origin. The transformations to be used are the following

$$\frac{d\theta}{dt} = \frac{1}{C^2} \qquad x = \xi\, C(t)$$
$$\psi(x,\, t) = C^{-1/2} \left[\exp \frac{i}{2} \frac{1}{C} \frac{dC}{dt} x^2\right] \phi(\xi,\, \theta) \quad ; \tag{22}$$

in |22| we rescale ψ but also introduce the $\exp \frac{i}{2} \frac{1}{C} \frac{dC}{dt} x^2$ term. Consequently we can show that if ψ satisfy

$$i \frac{\partial\psi}{\partial t} = - \frac{1}{2} \frac{\partial^2\psi}{\partial x^2} + V\,\psi \ , \tag{23}$$

we get the new Schroedinger equation

$$i \frac{\partial\phi}{\partial\theta} = - \frac{1}{2} \frac{\partial^2\phi}{\partial\xi^2} + \overline{V}\,\phi \tag{24}$$

with

$$\overline{V} = VC^2 + \frac{1}{2} C^3 \frac{d^2C}{dt^2}\, \xi^2 \ . \tag{25}$$

|25| is nothing else but the last part of equation |13|. The possibility to leave the Schroedinger equation quasi invariant has been used in (5) to solve a problem with time varying boundary conditions and, more precisely, the problèm of a particle trapped in a box the length of which is time dependent. Moreover the problem of multidimensional quantum harmonic oscillator with frequencies varying with time can with the same technique be brought back to problem of free particles. See (6).

VIII. Quasi Invariance and Critical Asymptotic Behaviour

We consider, finally, the motion of a point mass m around a center of force creating a time varying potential $V(t) = - \mu(t)/r$. The Hamiltonian is

$$H = \frac{1}{2} m\, V^2 - \frac{\mu(t)}{r}$$

in polar coordinates r, θ. The velocity is given by $V^2 = (\frac{dr}{dt})^2 + r^2 (\frac{d\theta}{dt})^2$

$$\frac{d^2r}{dt^2} = \frac{\sigma^2}{r^3} - \frac{\mu(t)}{m\, r^2} \qquad \text{with } \sigma = r^2 \left.\frac{d\theta}{dt}\right|_{t\,=\,0} \ . \tag{26}$$

We want to examine the asymptotic properties of |26| when the attractive potential $- \mu(t)/r$ – behaves as $\mu_o(1 + \Omega t)^{-n}/r$ (with n a positive real number and Ω

a characteristic frequency). We consider the usual transformation as given by |13|

$$\xi = r/C(t) \qquad d\theta = dt/C^2 \ . \qquad |27|$$

Substitution of |27| in |26| gives

$$\frac{d^2\xi}{d\theta^2} = \frac{\sigma^2}{\xi^3} - \frac{\lambda(\theta)}{m\,\xi^2} - \xi\, C^3 \frac{d^2C}{dt^2} \qquad |28|$$

with

$$\lambda(\theta) = C(t)\ \mu(t) \ .$$

We show how on this example we may now choose the arbitrary function C(t) to get some information about the asymptotic solution. The time behaviour of $\mu(t)$ suggests that we take

$$C(t) = (1 + \Omega t)^{\nu} \ . \qquad |29|$$

The second part of equation |27| and |29| combine to give

$$\Omega\theta = \frac{1}{1 - 2\nu}\left[(1 + \Omega t)^{1-2\nu} - 1\right] \quad \text{provided} \quad \nu < 1/2 \ . \qquad |30|$$

The problem is now the selection of ν. Table 1 indicates how ν is selected according the values of n to extract the asymptotic behaviour of the solution.

Table 1. Choice of the transformation parameter ν with respect to the mass index n

mass index n	Transformation index ν	new time θ	new mass λ	Remarks	
0	0	t	μ_o	static	
$0<n<\frac{1}{2}$	n	$\frac{(1+\Omega t)^{1-2n}-1}{\Omega\,(1-2n)}$	μ_o		
$n = \frac{1}{2}$	$\frac{1}{2}$	$\frac{1}{\Omega}\log(1+\Omega t)$	μ_o	critical (invariance)	logarithmic compression
$\frac{1}{2}<n<1$	$\frac{1}{2}$	$\frac{1}{\Omega}\log(1+\Omega t)$	$\mu_o \exp -\Omega\theta(n-\frac{1}{2})$		logarithmic compression
n = 1	1	$t/(1+\Omega t)$	μ_o	critical	renormalisation
1 < n	1	$t/(1+\Omega t)$	$\mu_o(1-\Omega\theta)^{n-1}$		renormalisation

We see two critical cases:

→ n = 1/2. The transformed problem involves a time invariant mass $\lambda(\theta) = \mu_o$ together with an external time independent transformation field (since in that case $C^3 \frac{d^2C}{dt^2}\xi = -\frac{\Omega^2}{4}\xi$). The problem is known to be integrable (7). Indeed if

we try on |26| the stretching transformation, $t = a^{\alpha}\,\bar{t}$ $r = a^{\beta}\,\bar{r}$

(and changing $(1+\Omega t)^{-n}$ in $(\frac{t}{T})^{-n}$) we see that the strict invariance is possible only if $\beta - 2\alpha = -3\beta - n\alpha$. But such an equation has non-trivial solutions only for $n = 1/2$.

→ n = 1 is the second critical case. The problem is equivalent to a constant mass with no transformation field and the time is renormalised.

These two cases are the only ones where exact integration can be obtained for all values of time. In the other cases only the asymptotic solutions can be obtained. Notice that if the cases $1/2 < n < 1$ and $1 < n$ correspond to two mathematically rather different treatments (with respective logarithmic time compression and time renormalisation) the interesting critical value from a physical point of view is $n = 1/2$. It has been shown (5) that for $n > 1/2$ the particle after a certain number of rotations will have a free particle trajectory with a constant velocity. It is of course obvious in the case $n > 1$ where renormalisation is possible but it is also true for $1/2 < n < 1$. For $n < 1/2$ on the other hand we can get an elliptical trajectory in the ξ plane which corresponds in r to spiralling particles.

IX. Conclusion

In this paper we have been specially interested in time evolution equations for models described either by a differential equation or by a partial derivative equation. Although we did not give an example a large amount of work has been devoted to evolution equations in phase space such as the Vlasov Poisson system found in plasma physics and beam dynamics.

Two techniques of utilisation are possible. In the first we have the equations strictly invariant and altough other transformations, more general than the stretching one, are possible (8) only these lead to explicit solutions. The price to pay is a poorer, more restricted solution, but the technique is especially useful when we start with a great number of variables and proceed to successive compaction of variables. Very often the "skeleton solution" left presents nevertheless a special physical interest. Notice finally that this technique is also quite useful to obtain the scaling laws of a physical system and allows developing from each existing "machine" a family of possible extensions. See (3) and (4) for the application to fusion machines.

The second technique is a little bit fuzzy and can be given an exact meaning only for systems which can be described by a classical or quantum Hamiltonian. In that case a generalised canonical transformation can be written which leads to the important concepts of time compression and time renormalisation. With this quasi-invariance technique we can often study the stability property of some of the solutions obtained with the stretching groups. Consequently the two techniques nicely compliment each other. This work can be extended in manydirections. First we can look at other fields such as hydrodynamics, general relativity, neutron transport theory and some encouraging results have already been obtained (9). Secondly the quasi-invariance technique can be combined with numerical analysis study to provide intelligent numerical integration of some of our equations (like the heat equation). It may be in this domain that the most interesting results will be found.

References

(1) L.I. Sedov, Similarity and dimensional methods, 4th Edition, ACADEMIC PRESS, 1959.

(2) G. Toulouse et P. Pfeuty, Introduction au groupe de renormalisation et à ses applications, Phènomène critique de transition de phase et autres, Presses Universitaires, Grenoble, 1975.

(3) J.R. Burgan, M.R. Feix, E. Fijalkow, A. Munier, R. Nakach, Journal de Physique, Colloque C3, Tome 41, pages 377-382, Avril 1980. See M.P. Moraux Thèse de 3ème Cycle, Université d'Orléans, Juin 1980.

(4) J. Gutierrez, A. Munier, J.R. Burgan, M.R. Feix and E. Fijalkow, in Nonlinear Problems in Theoretical Physics, Proc. 9th G.I.F.T. Seminar, June 1978, ed by A.F. Rañada, Lecture Notes in Physics, Vol. 98 (Springer Berlin, Heidelberg, New York 1979)

(5) J.R. Burgan, J. Gutierrez, M.R. Feix, E. Fijalkow and A. Munier, Applied Inverse Problems, ed. by P.S. Sabatier (Springer Berlin, Heidelberg, New York 1978)

(6) J.R. Burgan, M.R. Feix, E. Fijalkow and A. Munier, Physics Letters 74 A, 11-14, 1979.

(7) Vinti, M.N.R.A.S., 169, 417, 1974.

(8) Eisenhart L.P., Continuous group of transformation, Dover N.Y. 1961.

(9) A. Munier, J.R. Burgan, M.R. Feix and E. Fijalkow, Astrophysical Journal 236, 970, 1980.

Antecedent Invariant Curves of an Endomorphism. Influence Domain of a Stable Cycle Coexisting with an Isolated Stable Invariant Curve

R.L. Clerc and Ch. Hartmann

Université Paul Sabatier, U.E.R. Mathématiques, Informatique et Gestion, 118, Route de Narbonne, F-31062 Toulouse Cédex, France

Antecedent invariant curves of endomorphisms are more singular than those of diffeomorphisms because they are usually composed of the union of different antecedent branches. Their determination is essential for the study of the boundary of the total influence domain of an attractor. When two or more attractors coexist, antecedent invariant curves play also a key role in bifurcation involving the disappearance of one of these attractors.

Introduction

The singularity structure of the point mapping T of $\mathbb{R}^2$ onto $\mathbb{R}^2$

$$(x,y) \to (y, y-cx+x^2) \quad , \quad c \in \mathbb{R}^+ \tag{1}$$

is already known [1,2]. One of the singularities of (1) is an isolated closed invariant curve $\mathscr{C}$, which undergoes bifurcations at the parameter values $c = c_i^7$, $i = 1,2$, related to the existence of a cycle pair stable node-saddle, of order $k = 7$ [3]. In this paper antecedent invariant curves crossing saddles of T are considered. These curves allow describing the evolution of the influence domain of stable cycles of T as c traverses diverse bifurcation values. When c traverses $c = c_i^7$ (defined in [3]) T admits a non-classical stochastic bifurcation.

Determination of Antecedent Invariant Curves

It is obvious by inspection that the endomorphism (1) possesses two inverse branches T_ε^{-1}

$$(x,y) \to \left(\frac{c}{2}+\frac{\varepsilon}{2}\,[c^2+4(y-x)]^{1/2}, x\right) \quad , \quad \varepsilon = \pm 1 \quad . \tag{2}$$

It is this non-uniqueness, characteristic of non-conservative mappings, which complicates the determination of antecedent invariant curves crossing cycle points of T.

---

* Dynamic Systems Research Group, U.E.R.M.I.G., University Paul Sabatier, 118, Route de Narbonne, 31002 Toulouse, France

Given a cycle of order k of T with non-vanishing eigenvalues there exists for each point P_i (i=1,2,...,k) of this cycle one, and only one mapping

$$\Phi_i = T^{-1}_{\varepsilon_{i+k}} \circ T^{-1}_{\varepsilon_{i+k-1}} \circ \ldots \circ T^{-1}_{\varepsilon_i} \ , \quad \varepsilon_i = \pm 1 \ , \tag{3}$$

such that

$$\Phi_i(P_i) = P_i \quad .$$

The set of Φ_i is defined for $i = 1,2,\ldots,k$ by a circular permutation of the ε_i in (3).

<u>Definition</u>

Let the antecedent invariant curve $\mathscr{L}^-(P_i)$ traversing a point P_i of a cycle of order k of T be the analytical continuation of the Lattes arc (see [4] and [2] for a definition) if such an arc exists, subject to Φ_i.

It should be observed that $T^k[\mathscr{L}^-(P_i)] = \mathscr{L}^-(P_i)$. Let us recall that the local existence theorem [4] of an analytic consequent arc traversing a cycle point is valid whenever the eigenvalues of the cycle are $s \neq 0$ and $s \neq \pm 1$.

By convention (see for example [2,3]) the locus of points defined by $J(x,y) = 0$, where J is the Jacobian determinant of T, is called a critical curve of T of rank 0, and denoted LC_0. Similarly $LC_p = T^p(LC_0)$, i.e., the pth iterate of LC_0, is called a critical curve of T of rank p.

The mappings Φ_i have definition domains, whose boundaries belong necessarily to a union of critical curves T, $LC_1,\ldots,LC_k$ and of the antecedents of the latter up to the rank k (i.e., all iterates of LC_i, i=1,...,k by the set of mappings $(T_\varepsilon^{-1})^i$, i= 1,...,k, $\varepsilon=\pm 1$).

If $\mathscr{D}_i$ is the definition domain of the mapping Φ_i, then it is easily seen that an antecedent invariant curve $\mathscr{L}^-(P_i)$ belongs to the set

$$\mathscr{D}_i \cup \Phi_i(\mathscr{D}_i) \quad .$$

Hence, an invariant curve $\mathscr{L}^-(P_i)$ of T cannot "terminate" except at a critical curve $LC_1,\ldots,LC_k$ or one of its antecedents of rank 1 to 2k, or at an unstable invariant subvariety of T.

A numerical algorithm was constructed to permit the computation of antecedent invariant curves $\mathscr{L}^-(P_i)$ of endomorphisms like (1), when the cycle point of T is a saddle of order k (with eigenvalues $\neq 0$, +1, -1). By analogy with consequent invariant curves denoted by $\mathscr{L}_k$ in [2] and [3], the antecedent invariant crossing a saddle will be denoted by $\mathscr{L}_k^-$.

Determination of Influence Domains

Using the antecedent curves $\mathscr{L}_1^-$ crossing the saddle (c,c) of order one, the total influence domain of all attractors of T was determined. The set of $\mathscr{L}_1^-$ and of all its antecedents was found to be a closed bounded region D of $\mathbf{R}^2$, cut symmetrically by the curve LC_0 when $c \in [0,\bar{c}[$ (Fig.1). It should be noted that the curve LC_1 has two transverse intersections with the boundary of D. When $c = \bar{c} = 1.55$ the region D is still closed, LC_1 has now three intersections with the boundary of D, but only one of them is transverse (Fig.2). Finally, when $c > \bar{c}$ the region D is no longer closed, the "composite" attractor no longer exists, and LC_1 has only one transverse intersection with the set of antecedent invariant curves originated at (c,c). These invariant curves approach asymptotically the unstable focus (0,0) and its "excess" antecedent $T_{+1}^{-1}(0,0)$, i.e., its symmetrical image with respect to LC_0 (Fig.3).

The construction of $\mathscr{L}_7^-$ permits to study the evolution of the influence domain of the stable cycle of order k = 7, and by complementarity with respect to D the in-

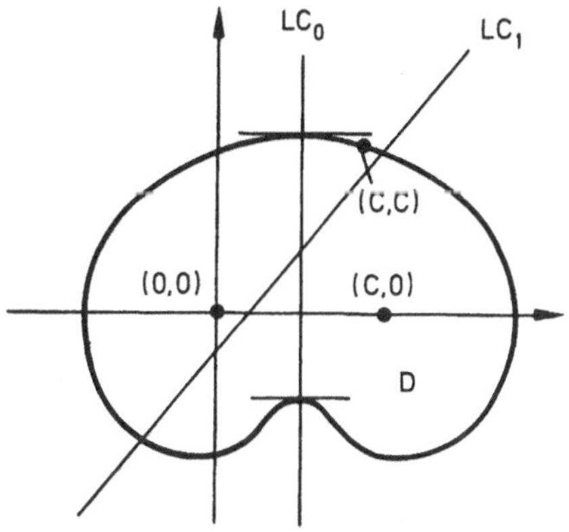

Fig. 1. Singly connected closed influence domain of (1), $0 < c < \bar{c}$. Two transverse intersections with LC_1

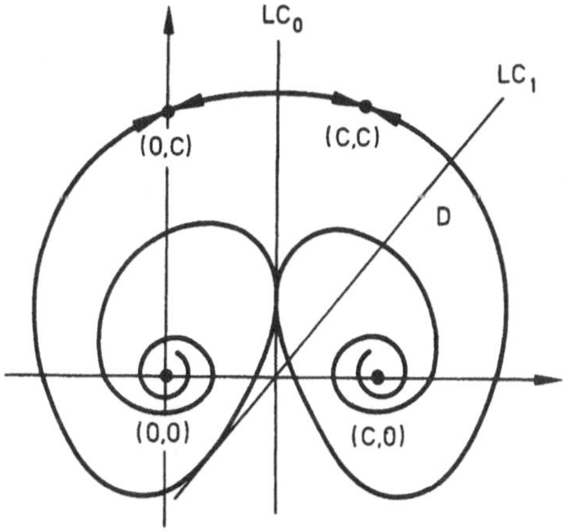

Fig. 2. Same as Fig.1, except $c = \bar{c}$. One transverse intersection persists, the second degenerates into a first order contact, and a third one appears

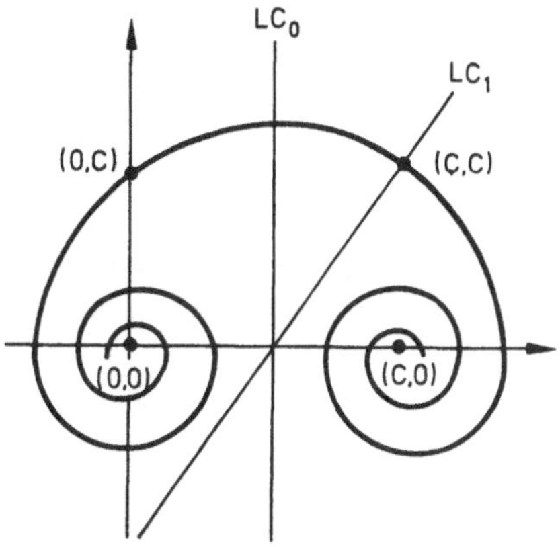

Fig. 3. Further evolution of Fig.1, $c > \bar{c}$. Only one transverse intersection with LC_1 remains. The singly connected closed influence domain no longer exists

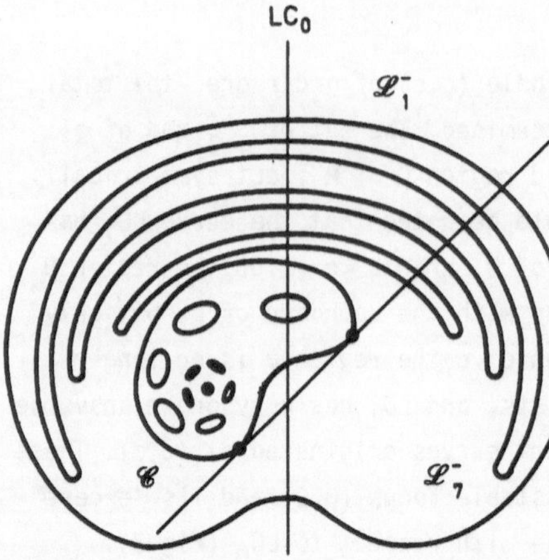

Fig. 4. Closed singly connected influence domain of the stable cycle of order $k = 7$ and some elements of the set of disjoint influence domains of one cycle point as seen through T_{+1}^{-1}; $c_1 < c < c_1^f$

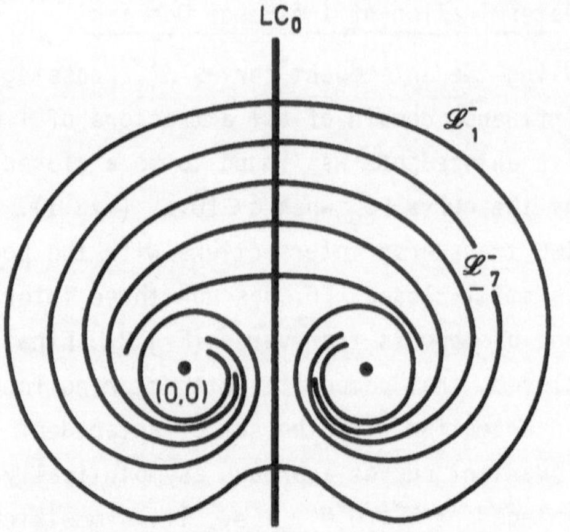

Fig. 5. The disjoint influence domains of Fig.4 have opened, forming spiral sectors inside the total influence domain D; $c_1^f < c < c_2^f$

fluence domain of the stable invariant curve $\mathscr{C}$. In the existence interval $[c_1, c_2]$ of the cycle pair $k = 7$ two values c_1^f, c_2^f can be found such that

$$c_1 < c_1^7 < c_1^f < c_2^f < c_2^7 < c_2 \quad .$$

When $c \in [c_1, c_1^f[$ the immediate influence domains of the points of the stable cycle $k = 7$ are closed, whereas when $c \in]c_1^f, c_2^f[$ they are open (inside D) and defined by spirals which approach asymptotically the unstable focus (0,0) and its excess antecedent (Fig.4 and 5, respectively). Of course, the immediate influence domains of the points of the stable cycle $k = 7$, as well as all antecedents of these domains are located inside D. It can be seen from Fig.4 that the point of the stable cycle $k = 7$, located inside the invariant curve $\mathscr{C}$, has a closed immediate influence domain, which together with its T_{+1}^{-1} antecedents of all orders forms a spiral chain composed of monotonically decreasing closed regions approaching asymptotically the unstable focus (0,0), and all links of this chain are also located inside $\mathscr{C}$.

Bifurcation Leading to the Disappearance of $\mathscr{C}$

The determination of the singular invariant curves $\mathscr{L}_k$ and $\mathscr{L}_k^-$, $k = 7$, makes it possible to show that the bifurcation destabilizing the isolated closed invariant curve $\mathscr{C}$ when c traverses the value c_1^7 (and the analogous value c_2^7) is a non-classical stochastic bifurcation. When $c = c_1^7 - \varepsilon$, $0 < \varepsilon << 1$ it was already observed (cf.[3]) that $\mathscr{C}$ is strongly oscillating near the points of the two cycles $k = 7$. The saddles $k = 7$ approach $\mathscr{C}$ but never attain it, because inside the c-interval in question the eigenvalues of the saddles are strictly different from unity. The local deformations of $\mathscr{C}$ are merely related to the proximity of the saddles.

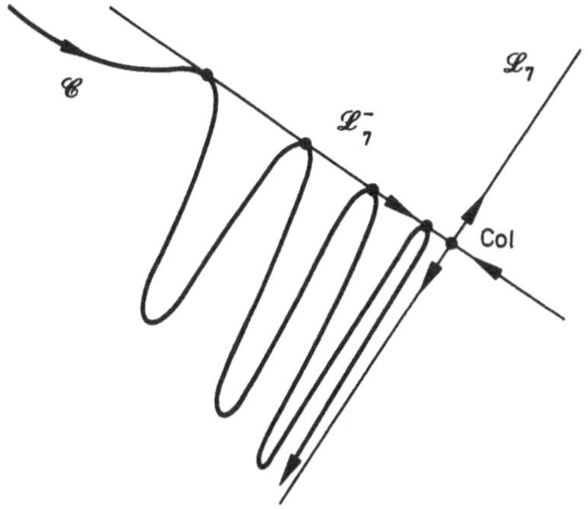

Fig. 6. Oscillations of the closed invariant curve $\mathscr{C}$ near a saddle of order $k=7$. $\mathscr{L}_7$, $\mathscr{L}_7^-$: consequent and antecedent invariant curves crossing the saddle; $c \lesssim c_1^7$

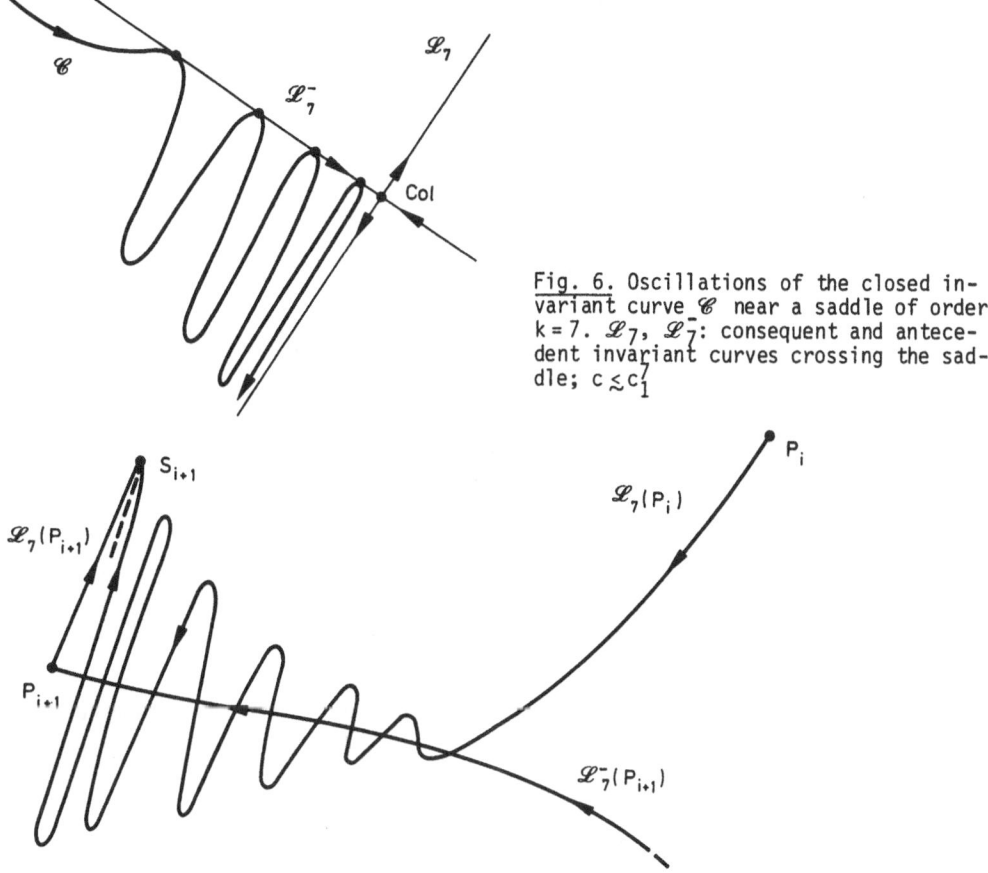

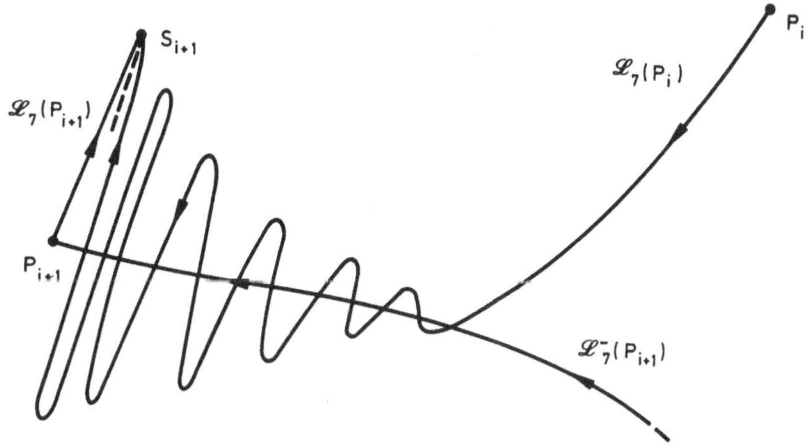

Fig. 7. Invariant curve bifurcation when the closed invariant curve $\mathscr{C}$ no longer exists; $c = c_1^7 + \varepsilon$, $\varepsilon > 0$, $\varepsilon \to 0$. Existence of homoclinic points on $\mathscr{L}_7$ and $\mathscr{L}_7^-$

When $\varepsilon \to 0$ the number of oscillations of $\mathscr{C}$ increases without limit near each saddle $k=7$, and the available numerical evidence points to the existence of heteroclinic points: the local maxima of the oscillatory parts of $\mathscr{C}$ appear to become tangent to an antecedent invariant curve $\mathscr{L}_7^-$ (Fig.6). In other words, $\mathscr{C}$ ceases to be smooth (of bounded variation) when $c \geq c_1^7 - \varepsilon$, $\varepsilon > 0$. It loses its stability via a tangency bifurcation involving heteroclinic points. Each saddle $k=7$ becomes an accumulation point of the latter. Consider now the $c = c_1^7 + \varepsilon$, $\varepsilon > 0$, side of the bifurcation of $\mathscr{C}$ by decreasing c till the value c_1^7, corresponding to the emergence of $\mathscr{C}$, is reached. For this purpose the invariant curves $\mathscr{L}_7$ and $\mathscr{L}_7^-$ traversing the saddles $k=7$ are examined. This procedure is more reliable than that based on the evolution of $\mathscr{C}$, because, unlike $\mathscr{C}$, $\mathscr{L}_7$ and $\mathscr{L}_7^-$ are continuations of analytically known curve segments, and not a result of purely numerical computations. Let P_i and P_{i+1} be two neighbouring saddles of order $k=7$. Since the rotation number of the cycle pair $k=7$ is unity, P_{i+1} is a consequent of P_i. The consequent invariant curves $\mathscr{L}_7$ initialized

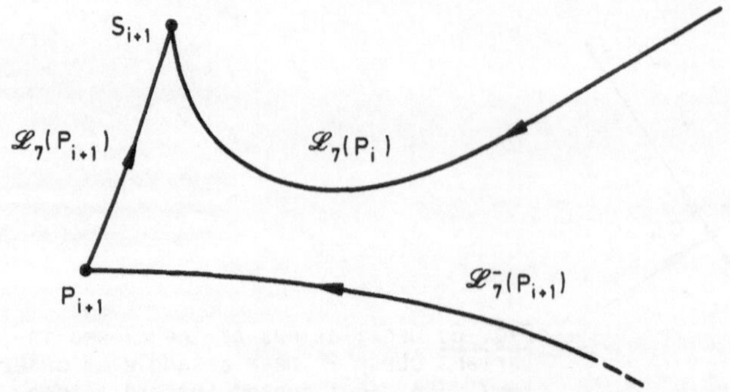

Fig. 8. Invariant curve configuration of Fig.7 after the tangency bifurcation leading to the disappearance of homoclinic points. $\mathscr{L}_7$ terminates at a stable cycle point; $c_1^7 < c_1 < c_2^7$

at P_i are found to intersect the antecedent invariant curves $\mathscr{L}_7^-$ initialized at P_{i+1}, producing thus an infinity of homoclinic points near each saddle $k=7$ (Fig.7). This situation is entirely consistent with that already reported on [3] for $c_1^7 < c < c_2^7$, with c not too close to c_1^7 (Fig.8). In the neighbourhood of P_{i+1} the invariant curve $\mathscr{L}_7(P_i)$ is a sort of envelope of the maxima of the oscillations shown in Fig.7. According to Birkhoff there exists thus an infinity of saddles of an arbitrarily high order near each homoclinic point, and a fortiori near each saddle $k=7$. The neighbourhood of the $k=7$ saddles is therefore stochastic. It was observed that the homoclinic points in question exist for $c \in]c_1^7, c_1^h[$, and disappear for $c = c_1^h$ when the invariant curves $\mathscr{L}_7^-(P_{i+1})$ and $\mathscr{L}_7(P_i)$ become tangent to each other ($c_1 < c_1^7 < c_1^f < c_1^h < c_2^7$). For $c = c_1^7 + \varepsilon$, $\varepsilon \to 0$, the stochastic bifurcation is similar to that observed in almost conservative mappings, but for $c = c_1^7 - \varepsilon$, $\varepsilon \to 0$, the evolution of the invariant curves appears to be novel. Whether the resulting heteroclinic points are stochastic or not, i.e., whether they imply the existence of an infinity of unstable cycles in their neighbourhood, is still an open problem. Another open problem concerns the evolution of an isolated closed invariant curve $\mathscr{C}$, generated at the destabilization of a focus by a Poincaré-Hopf bifurcation, as a function of the parametric distance $c - c_0$, c_0 being the bifurcation value. Does $\mathscr{C}$ lose its smoothness gradually or suddenly as $c - c_0$ increases? What method should be used to determine $\mathscr{C}$?

Conclusion

The study of antecedent and consequent invariant curves initialized at saddles has permitted the identification of a non-classical bifurcation leading to the disappearance of an isolated asymptotically stable closed invariant curve of an endomorphism.

References

1. R.L. Clerc, Ch. Hartmann: I.C.N.O., Prague, Tchécoslovaquie, Proceedings (1978) p.199
2. R.L. Clerc, Ch. Hartmann: C.R.Ac.Sc., Paris, t.289, A (1979) p.31
3. R.L. Clerc, Ch. Hartmann: C.R.Ac.Sc., Paris, mai 1980
4. S. Lattes: Ann. die Mathematica, (3), 13 (1906) pp.1-137

Topological Entropy As a Measure of Dynamic Chaos in Endomorphisms

J. Couot and I. Gumowski

Université Paul Sabatier, U.E.R. Mathématiques, Informatique et Gestion, 118, Route de Narbonne, F-31062 Toulouse Cédex, France

C. Gillot and C. Mira

I.N.S.A., Avenue de Rangueil, F-31077 Toulouse Cédex, France

Contemporary studies of endomorphisms are carried out as a rule from two different points of view. In the first, an endomorphism is considered as an implicit definition of a function, whose characteristic singularities (fixed points, cycles, invariant curves, etc.) as well as their qualitative properties (stability, instability, local phase space portrait, etc.) are sought directly or indirectly. In the second, an endomorphism is considered as a source of essentially geometric invariants, for example, invariant measures and various entropies. More particularly the topological entropy has been proposed as a measure of the amount of disorder or chaos contained in the endomorphism. It is illustrated by means of an example that a state of geometric chaos, characterized by a strictly positive value of the topological entropy does not necessarily coincide with a state of dynamic chaos.

The point of view that an endomorphism (or a diffeomorphism) defines implicitly a function of a discrete independent variable and of some continuous parameters (for example, initial values) is a consequence of the fact that an endomorphism (or a diffeomorphism) represents a mathematical formulation of a "natural" (i.e., physical, biological, etc.) evolution process, and that the independent variable represents discrete values of time. Since physical time has a preferential direction from the past to the future, the notion of stability and instability of stationary states (described by fixed points, cycles, certain invariant curves, etc.) plays a fundamental role, and in particular, asymptotically stable stationary states appear as physically more important than unstable ones. In fact, instability is viewed as either an entirely negative property (lack of stability) or only as conditionally positive one (defining point sets on the boundary of the influence domain of stable states).

The point of view that an endomorphism (or a diffeomorphism) is merely a source of geometric invariants implies that any differences between (physically) stable and unstable stationary states are either disregarded entirely, or that the notion of stability is unrelated to any preferential time arrow. For example, the distinction between stable and unstable fixed points or cycles plays no role in the determination of topological entropy. As a consequence it may happen that an endomorphism (defined inside a phase space region X and) possessing a unique asymptotically stable cycle (whose influence domain is contained inside X), may have the same value of topological entropy as another endomorphism, related to it by a continuous conjugate transformation but possessing only unstable cycles.

As an illustration of this property, consider the two endomorphisms T in $X = [0,1]$, $n = 0, \pm 1, \pm 2, \dots$:

$$x_{n+1} = 2(1-\mu)x_n + \mu \quad \text{if} \quad 0 \leq x_n \leq \frac{1}{2} \ ,$$
$$= 2(1-x_n) \quad \text{if} \quad \frac{1}{2} \leq x_n \leq 1 \ , \tag{1}$$

and

$$x_{n+1} = -\lambda^2 x_n^2 + 2\lambda(\lambda-1)x_n + \lambda(2-\lambda) \ . \tag{2}$$

When $\mu = 1/2$ in (1), see Fig.1, and $\lambda = \lambda_3 = 1.754\ 877\ 666$ in (2), see Fig.2, the topological entropy has the same value [$h=\log(1+\sqrt{5})/2>0$].

The value $\mu = 1/2$ in (1) is a bifurcation value corresponding to:

i) the appearance of an unstable double cycle of order three (which splits into two distinct unstable cycles of the same order for $\mu = 1/2+\varepsilon$, $0<\varepsilon<<1$),

ii) the coexistence of an infinity of unstable cycles of all orders, and

iii) the absence of any asymptotically stable cycle. Hence all discrete (half-) trajectories (i.e., the number or point sequences $\{x_n\}$, $n \geq 0$, initialized on any given $x_0 \in X$ not coinciding with a cycle point or its antecedent) are chaotic in both a geometric and a dynamic sense.

The value $\lambda = \lambda_0$ in (2) is a bifurcation value of a slightly different type:

i) there exists an asymptotically stable cycle $\bar{x}$ of order three with the (discrete) eigenvalue $s = 0$,

ii) $\bar{x}$ has a finite (nonzero) immediate influence domain,

iii) $\bar{x}$ coexists with an unstable companion cycle of the same order and with an infinity of unstable cycles of other orders,

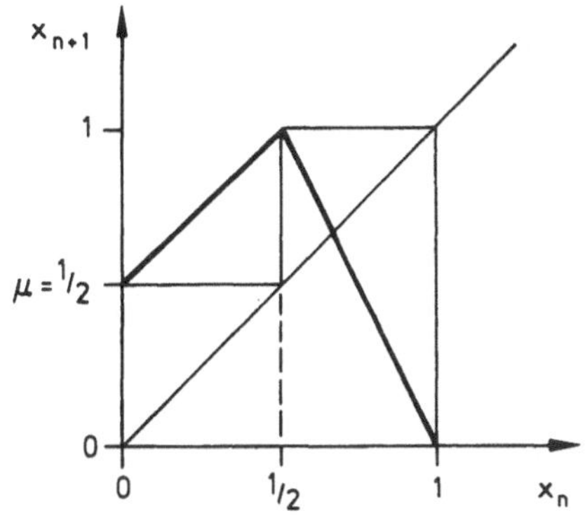

Fig. 1. x_{n+1} vs x_n, eq.(1), $\mu = 1/2$, same $h>0$. Geometric and dynamic "chaos"

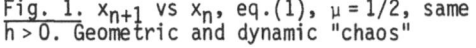

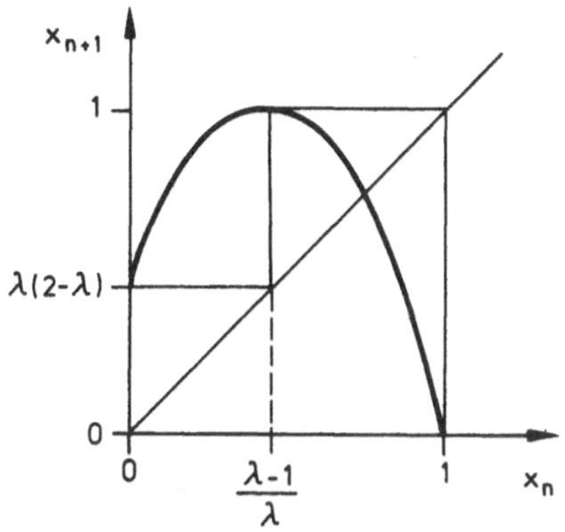

Fig. 2. x_{n+1} vs x_n, eq.(2), $\lambda = \lambda_0$, same $h>0$. Existence of an asymptotically stable cycle of order three. Geometric "chaos", dynamic "order"

iv) the interval X is a part of the total influence domain of $\bar{x}$. Except for the two cycles of order three (coincident in (1), split in (2)), all other cycles of (1) and (2) can be put into a one-to-one correspondence. Moreover these cycles have many common qualitative properties. For example, in addition to instability (diverging, initially close $\{x_n\}$, $n \geq 0$) they have the same rotation sequences [1]. In spite of these similarities the endomorphism (2) is not dynamically chaotic when $\lambda = \lambda_0$. In fact, for all x_0, $x_0 \in]0,1[$, not coinciding with a point of an unstable cycle (or its antecedent of a finite or infinite rank, including a limit point resulting from an accumulation of unstable cycles (cf.[1])), the points of any (half-) trajectory $\{x_n\}$, $n \geq 0$ will eventually enter the immediate influence domain of the asymptotically stable cycle $\bar{x}$, and then converge to the latter in a completely orderly manner. The set of the excluded x_0 is contained in the set of the boundary points of the total influence domain of $\bar{x}$, and appears thus dynamically as rather irrelevant (in spite of the fact that the number of exceptional x_0 is not finite). Expressed in another manner, this means that if one computes the $\{x_n\}$, $n \geq 0$ with a finite number of digits (or observes them with a finite accuracy, a physically unavoidable condition) then the exceptional set of x_0 behaves like it were qualitatively negligible ("all $\{x_n\}$, $n \geq 0$ finish a $\bar{x}$", i.e., $|\bar{x}-x_N| < \delta$ is attained with a finite N for an arbitrarily small physically meaningful value of the "resolution" $\delta > 0$).

Conclusion

A strictly nonzero value of topological entropy is a measure of geometric complexity (or chaos). It is not a sufficient condition for the existence of dynamic complexity (or chaos), unless the preferential time-arrow "past→present→future" is suitably taken into account. A sufficient condition for the breakdown of the equivalence between geometric and dynamic complexity is the existence of an asymptotically stable stationary state.

Reference

1. I. Gumowski, C. Mira: *Dynamique chaotique* (Cêpadues Editions, Toulouse 1980)

Topological Entropy of Markov Processes for a C^0- Endomorphism of the Interval

C. Gillot and G. Gillot

Groupe "Systèmes Dynamiques Non Linéaires et Applications", Laboratoire d'Analyse Numérique, Université Paul Sabatier Institut National des Sciences Appliquées, Avenue de Rangueil
F-31077 Toulouse Cédex, France

We study the topological entropy h of an one parameter endomorphism ψ of the unit interval, when its extremum trajectory is periodic. Some families of recurrent polynomials show the "local" behaviour of h, especially in the neighbourhood of the fixed point of ψ.

1 I is the interval $[0, 1]$. The non-invertible, C^0 map $\psi : I \to I$, is defined by

$$\psi(x) = 2(1-\zeta)\, x + \zeta \text{ for } x \in I_0 = [0, u] \qquad (1)$$

$$\psi(x) = 2(1-x) \text{ for } x \in I_1 = [u, 1], \qquad u = 1/2 .$$

ζ is a parameter such that $\zeta \in [0, 1[$. The trajectory of x_0, $Tr(x_0)$, is the set for the followers of x_0 :

$$Tr(x_0) = \{x_n \in I \mid x_n = \psi^n(x_0),\ x_0 \in I,\ n = 0, 1, 2, \ldots\} .$$

x_0 is a p-periodic point if there exists a positive integer p, (the smallest p) for which $x_p = x_0$.

We shall say that x_0 is eventually p-periodic if there exist a pair (m, p) of integers, $m \geqslant 0$, $p > 0$, such that x_m is p-periodic and x_i $(i = 0, 1, \ldots, m-1)$ is not periodic. The set $A(x_0) = \{x_0, \ldots, x_{m-1}\}$ is the transitory of x_0, its length $\ell(x_0) = \text{Card } A(x_0)$.

For $x_0 = u$, we have, by (1) : $Tr(u) = \{u, 1, 0, \zeta, \ldots\}$.

The aim of this work is the study of the topological entropy h, as a function of the parameter ζ, in the cases where the trajectory of u is periodic, or eventually periodic. More precisely ζ will take its values in the set τ :

$$\tau = \{\zeta \mid u \text{ is eventually periodic}, \ell(A(u)) \geqslant 0\}.$$

If $\zeta \in \tau$, the set Tr(u) contains only a finite number of distinct points ; these points, ranged in natural order, induct a finite partition E on I. Each segment $E_i \in E$ is mapped by ψ in some union of elements of E, so that the exchange process is typically a Markov's one. The topological entropy is then $h(\zeta) = \log \lambda$, where λ is the positive spectral radius of the transition matrix related to ψ_ζ, [4], [5].

Studying the map (1), C. MIRA [9] has exhibited the existence of the so-called "boîtes en file", limited by some values $\zeta^*_{2^i}$, i = 1, 2, ..., which are non classical bifurcations, and J. COUOT pointed, among other properties, that $h(\zeta) = 0$ for $\zeta > 3/4$, which is the cluster point of $\{\zeta^*_{2^i}\}$ [1].

2 In this section, we study the "local" behaviour of h, in the sense $\zeta \in \tau$, for some neighbourhood of the first non classical bifurcation $\zeta^*_{21} = 2/3$.

By definition (1), $\psi|I_1$ does not depend on ζ, and has the fixed point z = 2/3. If $\zeta = z$, we have $\psi(\zeta) = \zeta$, and $Tr(u) = \{u, 1, 0, \zeta, \zeta, \ldots\}$. The related characteristic equation is $f(\lambda) = \lambda^2 - 2$, so that the topological entropy is $h(2/3) = (\log 2)/2$. In this case the graph of ψ^2 on I is the duplication of that of ψ, for the value $\zeta = 0$.

Let ψ_1^{-1} be the inverse map of ψ/I_1. $\psi_1^{-1} : I \to I_1$ and $\psi_1^{-1}(x) = 1 - x/2$. Starting from u = 1/2, we define $\zeta_1 = \psi_1^{-1}(u)$, ... $\zeta_{n+1} = \psi_1^{-1}(\zeta_n)$. ζ_n is rational : $\zeta_1 = 3/4$, $\zeta_2 = 5/8$, ...

$$\zeta_n = (2^{n+2} + (-1)^{n+1})/3.2^{n+1} \quad . \tag{2}$$

The ζ_n whose indexes are of the same parity are on the same side of z. They verify

$$\zeta_{2n+1} - z = (z - \zeta_{2n})/2 \text{ , and}$$

$$\lim_{n \to \infty} \zeta_n = \zeta^*_{21} = 2/3. \tag{3}$$

Let us choose for ζ in (1) the value $\zeta = \zeta_{n-2}$, defined in (2). Then

$$Tr(u) = \{u, 1, 0, \zeta_{n-2}, \zeta_{n-3}, \ldots, \zeta_1, u, 1, 0, \ldots\}$$

so that u is (n+1)-periodic. We label these points in the following

$$x_1 = \zeta_{n-2} \text{ , } x_2 = \zeta_{n-3} \text{ , } \ldots, x_{n-1} = u, x_n = 1, x_{n+1} = 0$$

and we introduce the segments $E_1, \ldots, E_n$ such that

(i) E_1 has x_1, x_2 for endpoints

(ii) E_{i+1} has x_i, x_{i+2} for endpoints, i = 1, 2, ..., n-1.

Clearly

Lemma $\qquad \psi(E_1) = E_1 \cup E_2$

$$\psi(E_i) = E_{i+1} \text{ , } i = 2, 3, \ldots, n-1$$

$$\psi(E_n) = \begin{cases} E_2 \cup E_4 \ldots \cup E_{n-1} & \text{for n odd} \\ E_1 \cup E_3 \ldots \cup E_{n-1} & \text{for n even.} \end{cases}$$

The exchange property is thus Markov. $\mathcal{E}_n = [e_{ij}]$ is the n $\times$ n transition matrix of the process, where $e_{ij} = 1$ if $E_j \subset \psi(E_i)$, and 0 if not. The lemma yields

$$e_{11} = e_{12} = 1 \quad , \quad e_{1j} = 0 \quad , \quad j = 3, 4, \ldots, n$$

$$e_{ij} = \delta_{i+1}\text{ , }_j \quad , i = 2, 3, \ldots, n-1$$

$$e_{nj} = (1+(-1)^{n+j+1})/2 \quad , \quad j = 1, 2, \ldots, n \cdot$$

which give two types of $\mathcal{E}_n$ matrices, belonging to the parity of n.

If $P_n(\lambda)$ is the characteristic equation related to $\mathcal{E}_n$, one finds :

$P_{2n}(\lambda) = \lambda^{2n} + \lambda^{2n-1} - \lambda^{2n-2} + \lambda^{2n-3} + \dots - \lambda^2 + \lambda - 1.$

$P_{2n+1}(\lambda) = (1-\lambda)(\lambda^{2n} - \lambda^{2n-2} \dots - \lambda^2 - 1).$

These resume in the single form

$$P_n(\lambda) = \frac{(-1)^n \lambda^{n-1} (\lambda^2-2) - 1}{1 + \lambda} . \tag{4}$$

The polynomials P_n verify the recurrence relation

$$P_{n+1}(\lambda) = -\lambda P_n(\lambda) - 1 \qquad n = 2, 3, \dots \tag{5}$$

The case n = 2 corresponds to the value $\zeta = u = 1/2$, 3-periodic :

$P_2(\lambda) = \lambda^2 - \lambda - 1$, so that $h(1/2) = \log((1+\sqrt{5})/2)$.

We note θ_n the greatest positive root of P_n ($\mathcal{E}_n$ is irreducible). By (5), we are able to establish that the numbers θ_n are disposed, with respect to $\sqrt{2}$, in the following manner

$$\theta_3 = 1 < \theta_5 < \dots < \theta_{2n+1} < \dots < \sqrt{2} < \dots < \theta_{2n} < \dots < \theta_2 = (1+\sqrt{5})/2. \tag{6}$$

The value $\sqrt{2}$ is not a root of any P_n, for

$$P_n(\sqrt{2}) = 1 - \sqrt{2} \quad \text{for all } n \geqslant 2. \tag{7}$$

We can also localize each root θ_n. For instance

$$\sqrt{2} + 2^{-n-1} - n\, 2^{-2n-3/2} < \theta_{2n} < \sqrt{2} + 2^{-n-1}. \tag{8}$$

The indications obtained in (2), (3), (8) permit the evaluation of the quantity

$$\Delta h_{2n} = (\log \theta_{2n} - \log \theta_{2n+2}) \,/\, (\zeta_{2n} - \zeta_{2n+2})$$

which is an approximation of h derivative, if eventually this last exists. Nevertheless we have

$$|\Delta h_{2n}| > 2^n \quad \text{for all } n \geqslant n_o > 0. \tag{9}$$

The consideration of θ_{2n+1}, ζ_{2n+1} leads to a similar result. These observations are to be related to the ones done by J. Guckenheimer [6].

3 In view of extend the study in (2), we can imagine a set τ_1 of points ζ for which the u-trajectory contains the fixed point $z = 2/3$ of ψ, after a variable length transitory, i.e

$$\tau_1 = \{\zeta \,/\, z \in Tr(u)\}.$$

3.1 Here we are interested in the subset τ_1' of values ζ^p such that

$$Tr(u) = \{u, 1, 0, \zeta^p = y_1, y_2, \dots, y_{p-1}, z, z, \dots\} \tag{10}$$

with $0 < y_1 < y_2 < \dots < y_{p-1} < u$.

An adequate labelling of the intervals formed on I by the points of Tr(u) furnishes, for a fixed p, a relatively simple transition matrix, whose characteristic equation, with degree p + 2,

is $g(\lambda) = (-1)^p \lambda f_p(\lambda)$, and

$$f_p(\lambda) = \lambda^{p+1} - 2 \sum_{i=0}^{p-1} \lambda_i , \quad p \geqslant 1 \tag{11}$$

f_p is called "heart polynomial" of the situation ζ^p. If p = 1, we find $f_1(\lambda)=\lambda^2-2$. One observe the relation

$$f_{p+1}(\lambda) = \lambda f_p(\lambda) - 2$$

so that $f_p(2) = 2$ for all $p \geqslant 1$. Further, θ^p being the unique, larger than 1, root of $f_p(\lambda)$, one has

$$\theta^1 = \sqrt{2} < \theta^2 < \ldots < \theta^p < \ldots < 2. \tag{12}$$

Also the set τ_1 is infinite, and

$$\lim_{p \to \infty} \zeta^p = 0.$$

3.2 In this part, p is a fixed integer, p > 1. Taking back the method of section 2, we construct an infinite set of periodic values ζ_n^p defined by

$$Tr(u) = \{u, 1, 0, t_1, \ldots, t_{p-1}, x_1, x_2, \ldots, x_{n-1}, u, 1, 0, \ldots\}$$

where $0 < t_1 < t_2 < \ldots < t_{p-1}$, $\{t_i\} \subset I_o$

and $x_1 = \psi(t_{p-1})$, $\{x_j\} \subset I_1$, j = 1, 2, ..., n.

Which provides a partition of I by (n+p) intervals. The inverse maps of ψ being contractive, we deduce

$$\lim_{n \to \infty} \zeta_n^p = \zeta^p.$$

In other words, any neighbourhood of any point ζ^p contains, on the ζ - axis, an infinity of periodic points, whose periods are arbitrarily large.

As before, the $\mathcal{E}_n^p$ transition matrix has two forms. Its characteristic equation, rather delicate to compute, is given by

$$P_n^p(\lambda) = (-1)^p \frac{(-1)^n \lambda^n f_p(\lambda) + 1}{1 + \lambda}, \quad n \geqslant 1, \quad p \geqslant 1 \tag{13}$$

whence the "heart" name of f_p, for this term, appearing in (13) for all $n \geqslant 1$, describes the limit case in 3.1.

The P_n^p satisfy the recurrence relation

$$P_{n+1}^p(\lambda) = -\lambda P_n^p(\lambda) + (-1)^p. \tag{14}$$

Eq. (4) is a particular case of (13) : $P_{n+1}(\lambda)$ coïncide with $P_n^1(\lambda)$.

Also

$$P_n^p(\theta^p) = (-1)^p\ (1 + \theta^p)^{-1} \text{ for all } n \geqslant 1.$$

Again, θ_n^p being the largest positive root of P_n^p , the set $\{\theta_n^p\}$, bounded by θ^{p-1} and θ^{p+1}, is such that

$$\theta^{p-1} < \theta_2^p < \ldots < \theta_{2k}^p < \ldots < \theta^p < \ldots < \theta_{2k+1}^p < \ldots < \theta_1^p < \theta^{p+1} . \qquad (15)$$

4 The preceedings relations provide a very simple mean of computation of h, on a large number of points. Nevertheless, others more complexes sets are to be taken in consideration :

$$\tau_s = \{\zeta \ / \ u \text{ is eventually s-periodic}\} \qquad s = 1, 2, 3, \ldots$$

Note that, for a given s, this last definition does not fix any limitation for the length $\ell(u)$, neither the imbrication of its points. It is likely that the analytical study of such cases will not be simple.

We have elaborated a computational algorithm for values $h(\zeta)$, with $\zeta \in \tau_s$. It is based on the matrix transition notion, together with the u-trajectory. It is an efficient method, which is only limited by the exponential growth of location errors of the iterates of u : indeed this is the inherent phenomenon to the proper essence of such endomorphisms types.

The Figure 1 is drawn only with points of τ_1, τ_2, for values of ζ such that $\ell(u) < 12$. There are about 10^3 points. Figure 2 shows the results obtained in section 2, more specially (9) ; it contains some additional points of τ_1, τ_2, in the neighbourhood of $\zeta^*_{2^1}$.

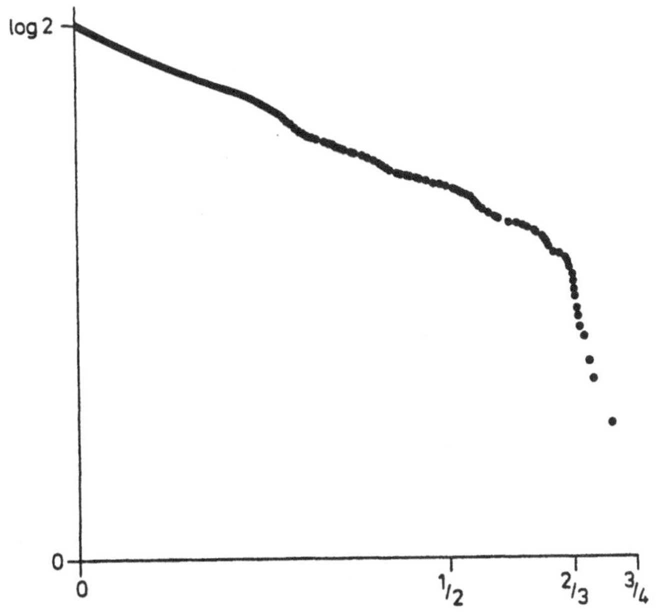

Fig.1.

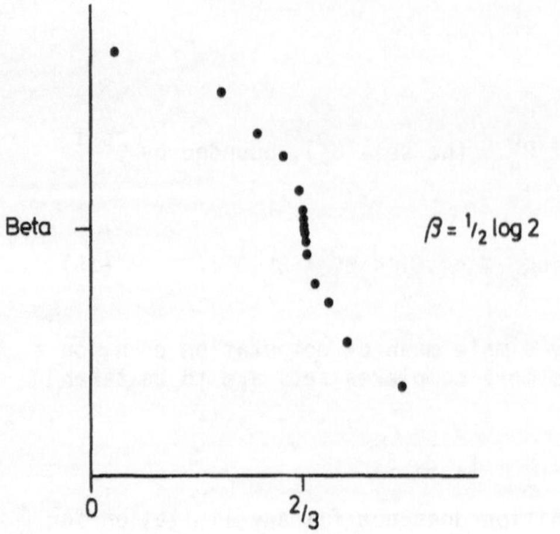

Fig.2.

Note, fig. 1, that all points of sets τ_1, τ_2 which have been computed are less than $\zeta^*_{2^2} = 0{,}7168 \ldots$. It is known that, for that second bifurcation value, the entropy is

$$h(\zeta^*_{2^2}) = (\log 2)/4.$$

In fact, the τ_1 points have already a rather complicated imbrication structure in terms of their u-trajectory, so that we have first restricted the analytical study to the subset τ_1', for in this case the description of Tr(u) is easy.

Note also that the graph 1, made with a drawing table, seems to be a continuous one, in the neighbourhood of 0. This is to be related to

$$\lim_{p \to \infty} \zeta^p = 0.$$

In view of obtaining a readable graph, we have not mentioned points of ζ_n^p type.

Concerning fig. 2, recall that (9) is uniquely established on points ζ_{2n}, ζ_{2n+1}, described in 2. But we do not know if a similar situation occurs for another ζ between 0 and $\zeta^*_{2^1}$

Conclusion This study shows the skeleton of topological entropy $h(\zeta)$, for periodic or eventually periodic values of ζ. The word "local behaviour" is to be understood in this sense.

Another interesting problem would be the knowledge of h, even on pointwise form, for non periodic ζ. In view of that question, another computational algorithm has been written, using the results of M. MISIUREWICZ, [10]. This last method, suitable for any ζ, evaluates the number of extrema appearing in ψ^n, $n = 1, 2, \ldots$, which necessarily leads to very large computation amounts, greater than those employed for the results [2], [3].

Finally, we think that a finer approach of the problems related to Tr(u), [7], [8], will imply a better observation of the entropy function.

References

1 J. COUOT. *"Invariant Measures and Topological Entropy of Complex Stationary States of Dynamic Systems..."* Proceedings of the VIIIth ICNO, 1978, Prague, p. 205-210.

2 J. COUOT, C. GILLOT, G. GILLOT. *"Détermination Numérique des Densités de Suites Récurrentes"*. Comportement des Processus Itératifs, 1979, La Garde Freinet.

3 J. COUOT, C. GILLOT, G. GILLOT. *"Quelques Simulations Numériques de Densités de Suites Récurrentes"*. Publications du Groupe "Systèmes Dynamiques non Linéaires", 1979.

4 M. DENKER, C. GRILLENBERGER, K. SIGMUND. *"Ergodic Theory on Compact Spaces"*, Lecture Notes in Mathematics, Vol. 527 (Springer Berlin, Heidelberg, New York 1976)

5 F.R. GANTMACHER. *"Théorie des Matrices"*. Dunod, 1966, Paris.

6 J. GUCKENHEIMER. *"Sensitive Dependance to Initial Conditions for One Dimensional Maps"*. Publications I.H.E.S., 1979.

7 L. JONKER. *"Periodic Orbits and Kneading Invariants"*. Proc. London. Math. Soc (3) 1979, p. 428-450.

8 I. MILNOR, W. THURSTON. *"On Iterated Maps of the Interval. The Kneading Matrix"*. Preprint.

9 C. MIRA. *"Dynamique Complexe engendrée par une Récurrence, ou Transformation Ponctuelle, Continue, Linéaire par Morceaux"*. C.R. Acad. Sci. Paris (285) 1977, p. 731-734.

10 M. MISIUREWICZ, W. SLENK. *"Entropy of Piecewise Monotone Mappings"*. Asterisque, Soc. Math. de France (50), 1977.

Sequential Iteration of Threshold Functions

E. Golès Chacc*

Laboratoire IMAG, B.P. 53X
F-38041 Grenoble Cédex, France

Abstract: Let Δ be a function from $\{0,1\}^n$ into itself, whose components are threshold functions. We study the convergence of sequential iteration on Δ. This includes the behaviour of "majority rule" in the spin glass problem.

1. The Main Theorem

Let $E = \{0,1\}$ and Δ be a function :

$$\Delta : E^n \to E^n$$

$$y = (y_1,\ldots,y_n) \to (\phi_1(y),\ldots,\phi_n(y))$$

where each ϕ_i is a threshold function, defined by :

$$\phi_i(y) = \begin{cases} 0 & \text{if} \quad \sum_{j=1}^{n} \alpha_{ij} y_j < \theta_i \\ 1 & \text{otherwise} . \end{cases}$$

We say that Δ is "symmetrically threshold" if for each pair (i,j) we have $\alpha_{ij} = \alpha_{ji}$.

We study the behaviour of the "sequentially iteration" :

$$\begin{aligned} y_1^{r+1} &= \phi_1(y_1^r,\ldots,y_n^r) \\ &\vdots \\ y_i^{r+1} &= \phi_i(y_1^{r+1},\ldots,y_{i-1}^{r+1},y_i^r,\ldots,y_n^r) \qquad , \qquad y^\circ \in E^n \\ &\vdots \\ y_n^{r+1} &= \phi_i(y_1^{r+1},\ldots,y_{n-1}^{r+1},y_n^r) . \end{aligned}$$

* Adresse à partir de Décembre 1980: Departamento Matematicas. Universidad de Chile; Casilla 5272, Correo 3, Santiago, Chile.

The main theorem is the following :

Theorem: *If* $\alpha_{ii} > 0$ *for* $i \in \{1,\dots,n\}$ *then* :

For any $y^\circ \in E^n$, *there exists* $s \in \mathbb{N}$, *such that* :

$y^{s+1} = y^s$ *(convergence to a fixed point).*

2. The Tools of Proof

Clearly, since E^n is finite for each $y^\circ \in E^n$ there exists s , $T \in \mathbb{N}$, $s \geq 0$ $T > 0$, such that :

$$y^{s+T} = y^s \quad \text{and} \quad y^{s+r} \neq y^s \quad \text{for any} \quad r \in \{0,\dots,T-1\} .$$

Then, consider the $n \times T$ matrix :

$$X(y^\circ,T) = \begin{pmatrix} x_1(0) & \dots & x_1(T-1) \\ \vdots & & \vdots \\ x_n(0) & \dots & x_n(T-1) \end{pmatrix} = (y^s, y^{s+1}, \dots, y^{s+T-1}) .$$

It is obvious that, for $i \in \{1,\dots,n\}$:

$$x_i(0) = \begin{cases} 0 & \text{if} \quad \sum_{j=1}^{i-1} \alpha_{ij}\, x_j(0) + \sum_{j=i}^{n} \alpha_{ij}\, x_j(T-1) < \theta_i \\ 1 & \text{otherwise} \end{cases}$$

$$x_i(\ell) = \begin{cases} 0 & \text{if} \quad \sum_{j=1}^{i-1} \alpha_{ij}\, x_j(\ell) + \sum_{j=i}^{n} \alpha_{ij}\, x_j(\ell-1) < \theta_i \\ 1 & \text{otherwise} \end{cases} .$$

$$\text{for} \quad \ell \in \{1,\dots,T-1\}$$

Let γ_i denote the smallest period of the row x_i (γ_i is necessarily a divisor of T). Let us define the mapping L on the row's set of $X(y^\circ,T)$:

$$L : S \times S \to \mathbb{R}$$

$$(x_i,x_j) \to L(x_i,x_j)$$

$$L(x_i,x_j) = \begin{cases} \alpha_{ij} \sum_{\ell=0}^{T-1} x_i(\ell)(x_j(\ell+1)-x_j(\ell)) & \text{if} \quad j < i \\ 0 & \text{if} \quad i = j \\ \alpha_{ij} \sum_{\ell=0}^{T-1} x_i(\ell)(x_j(\ell)-x_j(\ell-1)) & \text{if} \quad j > i \end{cases} .$$

<u>Lemma 1:</u>

(*i*) $L(x_i,x_j) + L(x_j,x_i) = 0$ *for* $i,j \in \{1,\dots,n\}$

(*ii*) *If* $\gamma_i = 1$, *then* $L(x_i,x_j) = 0$ *for* $j \in \{1,\dots,n\}$.

The proof follows easily from the symmetrical hypothesis and the definition of L .

<u>Lemma 2:</u> *Let us* $x_i \in S$ *such that* $\gamma_i \geq 2$ *and* $\alpha_{ii} \geq 0$, *then* :

$$\sum_{j=1}^{n} L(x_i,x_j) < 0 .$$

For the proof, rather technical, we use a partition of the index set $\{\ell \in \{0,\dots,T-1\} \mid x_i(\ell) = 1\}$ and the threshold values associate to each $x_i(\ell)$ [4].

The proof of the theorem follows easily, by the lemmas as before :

For $y^\circ \in E^n$, let $X(y^\circ,T)$ be the corresponding matrix. If $T \geq 2$, then at least one γ_i is greater or equal to 2 .

By lemmas (1,ii) and (2) we have :

$$\sum_{i=1}^{n}\sum_{j=1}^{n} L(x_i,x_j) < 0$$

but, by lemma (1,i) $\sum_{i=1}^{n}\sum_{j=1}^{n} L(x_i,x_j) = 0$, which is a contradiction.

3. Comments

The theorem , is not true in general, if Δ does not satisfy the symmetrical threshold or the sign condition on α_{ii} .

(*) If $\alpha_{ij} \neq \alpha_{ji}$ the following is a counter example :

$$\Delta : E^6 \to E^6$$

$$\phi_1(y) = \begin{cases} 0 & \text{if } y_3 + y_5 < 1 \\ 1 & \text{otherwise} \end{cases} ; \quad \phi_6(y) = \begin{cases} 0 & \text{if } y_1 < 1 \\ 1 & \text{otherwise} \end{cases} ;$$

$$\phi_j(y) = \begin{cases} 0 & \text{if} \quad y_{j+1} < 1 \\ 1 & \text{otherwise} \end{cases} \qquad \text{for} \quad j \in \{2,3,4,5\} .$$

We have a two length cycle :

$$(1,1,0,1,0,1) \rightleftarrows (0,0,1,0,1,0)$$

Fig. 1

(*) If the α_{ii} are not all non-negative, we can have cycles :

$$\Delta : E^4 \to E^4$$

$$\phi_1(y) = \begin{cases} 0 & \text{if} \quad -y_1+y_2-y_4 < 0 \\ 1 & \text{otherwise} \end{cases} \quad ; \quad \phi_2(y) = \begin{cases} 0 & \text{if} \quad y_1-y_2-y_3 < 0 \\ 1 & \text{otherwise} \end{cases}$$

$$\phi_3(y) = \begin{cases} 0 & \text{if} \quad -y_2-y_3+y_4 < 0 \\ 1 & \text{otherwise} \end{cases} \quad ; \quad \phi_4(y) = \begin{cases} 0 & \text{if} \quad -y_1+y_3-y_4 < 0 \\ 1 & \text{otherwise} \end{cases} .$$

We have a three length cycle :

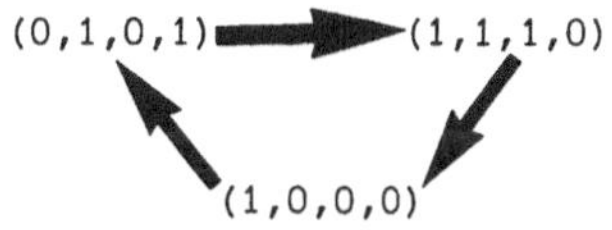

Fig. 2

4. Applications to Spin Glass Problem

The Ising Model with random nearest neighbor interactions represents a certain number of magnetic impurities (Fe atoms) in a non-magnetic metal [2]. If to each Fe atom we associate a single variable $y_i \in \{0,1\}$, indicating spin orientation, and to each pair (i,j) of neighbor spins a values $\alpha_{ij} \in \{-1,1\}$ indicating spin interaction, the physical problem is : *to find a spin configuration, such that minimising the number of unbalanced interactions*, where (i,j) is unbalanced if :

$y_i = y_j$ and $\alpha_{ij} = -1$

or

$y_i \neq y_j$ and $\alpha_{ij} = 1$.

D'Auriac, Robert, Maynard proposed a Monte-Carlo algorithm, for the local majority rule, to find a minimaly spin configuration [1,5]. In the present section we do a sequentially deterministic iteration, in a torus, and by the main theorem, a fixed point convergence condition (local minimum) is established.

Let us take a finite square lattice with periodic boundary condition (a torus) and interactions $\alpha_{ij} \in \{-1,1\}$:

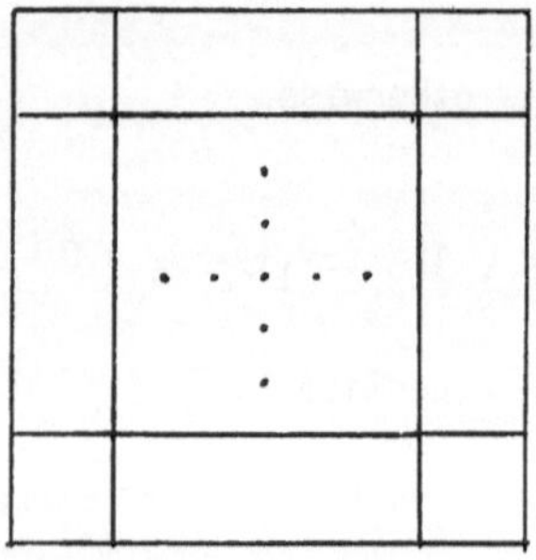

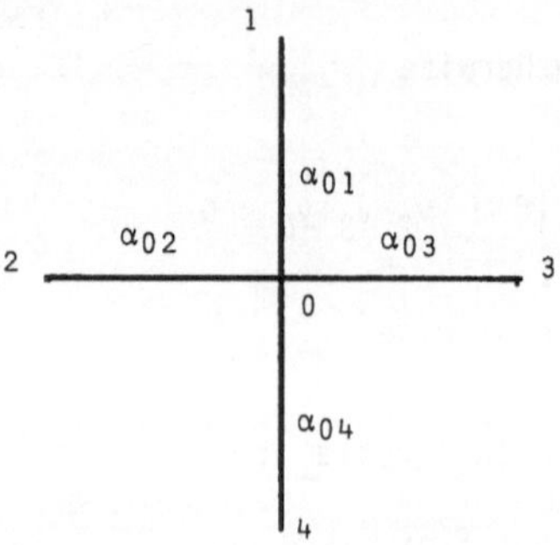

Fig. 3

The majority rules are the followings :

$$\text{A Maj}\,(y_0,y_1,y_2,y_3,y_4) = \begin{cases} 0 & \text{if} \quad \alpha y_0 + \sum_{j=1}^{4} c(y_j) < 3 \\ 1 & \text{otherwise} \end{cases}$$

$$\text{B Maj}\,(y_0,y_1,y_2,y_3,y_4) = \begin{cases} 0 & \text{if} \quad -\alpha y_0 + \sum_{j=1}^{4} c(y_j) < 2 \\ 1 & \text{otherwise} \end{cases}$$

with $\alpha \in [0,+\infty[$ and $c(y_j) = y_j$ if $\alpha_{oj} = +1$, $1-y_j$ otherwise .

If we do the sequentially iteration by rows, then :

* *If we use in each vertex the* A Maj. *rule, from the main theorem, we have always convergence to a fixed point (local minimum).*

* *If we use in each vertex the* B Maj *rule (crazy spin), the convergence to a fixed point it's not assured. We can found periodic spin configuration.*

<u>Examples</u> : (3 x 3 torus)

(a) α_{ij} = 1 for each pair (i,j) of neighbor spins.

```
0 + 1 + 0      0 + 0 + 0
+   +   +      +   +   +      A Maj rule
1 + 0 + 1  →   0 + 0 + 0      Fixed point
+   +   +      +   +   +
0 + 1 + 0      0 + 0 + 0

0 + 1 + 0      1 + 0 + 1
+   +   +  →   +   +   +
1 + 0 + 1      0 + 1 + 0      B Maj rule
+   +   +  ←   +   +   +      Two length cycle
0 + 1 + 0      1 + 0 + 1
```

<u>Fig. 4</u>

(b) α_{ij} = -1 for each pair (i,j) of neighbor spins.

```
0 - 1 - 0      0 - 1 - 0
-   -   -      -   -   -      A Maj rule
1 - 0 - 1  →   1 - 0 - 1      Fixed point
-   -   -      -   -   -
0 - 1 - 0      0 - 1 - 0

0 - 1 - 0      1 - 0 - 1
-   -   -  →   -   -   -      B Maj rule
1 - 0 - 1      0 - 1 - 0      Two length cycle
-   -   -  ←   -   -   -
0 - 1 - 0      1 - 0 - 1
```

<u>Fig. 5</u>

(c) $\alpha_{ij} = 1$ for each pair (i,j) of neighbor's row spins ; $\alpha_{ij} = -1$ otherwise.

```
0 + 1 + 0      0 + 0 + 0
-   -   -      -   -   -
1 + 0 + 1  →   1 + 1 + 1      A Maj rule
-   -   -      -   -   -      Fixed point
0 + 1 + 0      0 + 0 + 0

0 + 1 + 0      1 + 0 + 1
-   -   -  →   -   -   -
1 + 0 + 1      0 + 1 + 0      B Maj rule
-   -   -  ←   -   -   -      Two length cycle
0 + 1 + 0      1 + 0 + 1
```

Fig. 6

This behaviour (two length cycle for B Maj rule) is general for n x n torus, n odd. We conjecture that, for a n x n torus and $\alpha_{ij} \in \{-1,1\}$, the sequential iteration on B Maj rule, has at least a periodic spin configuration.

References

1 D'AURIAC ANGLES et VILLON P. "Fluctuations d'aimantation dans un verre de spin par simulation numérique de Monte-Carlo". Rapport D.E.A. Analyse Numérique, Grenoble 1978.

2 BARAHONA F. "Sur la complexité du problème du verre de spin". Rapport de Recherche n° 171, IMAG, 1979.

3 GOLES E. and OLIVOS J. "Periodic behaviour of generalized threshold functions", Discrete Mathematics, 30 (1980) 187-189.

4 GOLES E. "Etude des itérations dans un ensemble fini". Thèse Docteur-Ingénieur, Grenoble, 1980, à paraître.

5 ROBERT F. "Une approche booléenne du problème de la frustration". Séminaire Analyse Numérique, n° 302, IMAG, Grenoble 1978.

Some Properties of Second Order Dynamic Systems with Parametric Resonances

I. Gumowski

Université Paul Sabatier, U.E.R. Mathématiques, Informatique et Gestion, 118, Route de Narbonne, F-31062 Toulouse Cédex, France

The 1/2-subharmonic resonances of three particular dynamic systems are examined from the point of view of the theory of recurrences. It is found that the construction of approximate subharmonic solutions depends critically on the determination of characteristic exponents near bifurcations.

Subharmonic Resonances of Linear Systems

It has been observed both experimentally and theoretically that dynamic systems with periodically varying parameters possess subharmonic resonances. The simplest system is described by the Mathieu equation:

$$\ddot{x} + (a+h\cdot\cos 2t)x = 0 \quad , \quad a,\ h,\ t,\ x(t) = \text{real valued} \quad . \tag{1}$$

For some values of a and h, eq.(1) admits periodic solutions $x(t)$ of period 2π, although the "coefficient of elasticity" $(a+h \cos 2t)$ oscillates with a period π. The solutions of (1) can be expressed by means of the characteristic number μ, which plays the role of an eigenvalue in the Floquet functional equation:

$$x(t+\pi) = e^{2\pi\mu}\, x(t) \quad . \tag{2}$$

A subharmonic solution of (1) exists when $\mu = 0$ in (2). The linearity of (1) permits to reduce (2) to a purely algebraic form. If $\xi(t)$ is a particular solution of (1), satisfying the initial conditions $x(0) = 1$, $\dot{x}(0) = 0$, then:

$$\text{ch}(2\pi\mu) = \xi(\pi) \quad ,$$

or explicitly:

$$\mu = \frac{1}{2} \arg \text{ch}[\xi(\pi)] \quad . \tag{3}$$

Since it is possible to determine $\xi(\pi)$ by several methods (in the worst case, numerically) the determination of μ as a function of a and h is straightforward. The determination of the locus "h vs a" such that $\mu = 0$ is more tedious because it requires a root-finding technique.

In the case of the more general linear equation

$$\ddot{x} + p(t)\cdot\dot{x} + q(t)\cdot x = 0 \quad , \tag{4}$$

where p, q are at least piecewise continuous periodic functions of period $T > 0$, it is necessary to consider two particular solutions $\xi(t)$, $\eta(t)$ of (4), satisfying the initial conditions $x(0) = 1$, $\dot{x}(0) = 0$ and $x(0) = 0$, $\dot{x}(0) = 1$, respectively. The functional equation (2) implies then that $\lambda = e^{2\pi\mu}$ is a root of the characteristic equation:

$$\lambda^2 - 2A\lambda + W = 0 \quad , \quad 2A = \xi(T) + \dot{\eta}(T) \quad , \quad W = \exp\left(\int_0^T p(t)dt\right) \quad . \tag{5}$$

In order to determine the roots λ_1, λ_2 of (5), it is thus necessary to know two particular solutions of (4) inside the interval $0 \leq t \leq T$. It is easily verified that the above quadratic polynomial in λ is also the characteristic equation of the recurrence

$$y_{n+2} - 2Ay_{n+1} + Wy_n = 0 \quad , \quad n = 0, \pm 1, \ldots \quad , \tag{6}$$

where n represents time-instants separated by one period T. The subharmonic solution of period 2π of (1) corresponds to the critical case $\lambda_1 = \lambda_2 = 1$ of (6). The recurrence (6) admits, however, also other critical values of λ which may lead to a subharmonic resonance [1]. For example, when $W \neq 1$, eq.(4) admits a periodic solution of period 2T, provided $\lambda_1 = -1$, $|\lambda_2| \neq 1$. A typical example is

$$\ddot{x} + 2\alpha\dot{x} + \omega_0^2(1-h\cos 2\omega t)x = 0 \quad , \quad 0 < h < 1 \quad , \quad 0 < \alpha \ll 1 \quad . \tag{7}$$

When h is larger than the threshold value

$$h_0 = 2(b^2+4\alpha\omega^2)^{1/2}/\omega^2 \quad , \quad b = \omega_0 - \omega \quad , \tag{8}$$

then for $c = h - h_0 \ll 1$, there exists a one-parameter family of periodic solutions of period $T = 2\pi/\omega$ of (7) described by [2]

$$x(t) = C(A_1 \cos\omega t + B_1 \sin\omega t + A_3 \cos 3\omega t + B_3 \sin 3\omega t + \ldots) \quad , \tag{9}$$

where the coefficients A_i, B_i are known functions of c, and C is an arbitrary constant. The second linearly independent solution of (7) is damped:

$$x(t) = C_1 e^{-\gamma t}\cdot\varphi(t) \quad , \quad \varphi(t+T) = \varphi(t) \quad , \quad C_1 = \text{arbitrary constant} \quad . \tag{9a}$$

Subharmonic Resonances of Nonlinear Systems

The functional equation (2) cannot be used directly in a nonlinear context, except to study bifurcation solutions of a linear "generating" equation. A reduction of a differential equation to an equivalent recurrence is, however, still possible (for example, by means of Poincaré's method of sections). The cycles of order k of the recurrence represent then 1/k-subharmonic resonances of the differential equation. The determination of the equivalent recurrence is still straightforward: it is sufficient to establish an explicit relation between $x_0 = x(0)$, $y_0 = \dot{x}(0)$ and $x_1 = x(T)$, $y_1 = \dot{x}(T)$, which defines unambiguously the single-valued functions f and g in

$$x_{n+1} = g(x_n, y_n) \quad , \quad y_{n+1} = f(x_n, y_n) \quad , \quad n = 0, 1, \ldots \quad . \tag{10}$$

The practical procedure is unfortunately very laborious. Series expansions involve as a rule small denominators as well as secular terms, and numerical computations require a large number of particular solutions (at least in the proximity of bifurcations). At present it is more efficient to use rough estimates of f and g in qualitative arguments, allowing to establish the existence, uniqueness and type of a subharmonic resonance, instead of using highly precise estimates of these functions in an actual computation of a subharmonic solution. When $\lambda_1 = \lambda_2 = 1$ for a fixed point of (10), the subharmonic solution is called doubly degenerate, and when $\lambda_1 = -1$, $|\lambda_2| \neq 1$, it is called singly degenerate.

The simplest illustrative example is probably the quasi-linear equation [3]:

$$\ddot{x} + \omega_0^2(1-h\cos\omega t)x = -\varepsilon(2\alpha\dot{x}+\beta x^3) \quad , \quad h,\alpha,\beta > 0 \quad , \quad 0 < \varepsilon \ll 1 \quad . \tag{11}$$

The $2\omega_0 = \omega$ subharmonic resonance is for all practical purposes doubly degenerate, and it is possible to use near it an asymptotic method [4]:

$$x(t) \simeq a\cos(\omega_0 t+\theta_0) \quad , \quad a^2 \simeq \left[\omega^2-4\omega_0^2\pm2(h\omega_0^4-4\beta^2\omega^2)^{1/2}\right]\Big/(3\beta) \quad , \tag{12}$$

provided $h > h_0 = 4\alpha/\omega_0$; θ_0 is an integration constant.

A somewhat more complicated example is [2]

$$\ddot{x} + 2\alpha\dot{x} + \omega_0^2(1-h\cos\omega t)x = -\beta x^2 \quad , \quad h,\alpha,\beta > 0 \quad . \tag{13}$$

The subharmonic resonance $2\omega_0 = \omega$ is singly degenerate. The equivalent recurrence involves the bifurcation:

stable node $\rightarrow$ saddle + stable cycle node of order $k = 2$.

The threshold value h_0 of h and an approximate subharmonic solution can be found by at least two different methods (cf.[2] and [5]). If $\omega_0 = 1$, $\omega \simeq 2$, and $2\omega - (h-h_0) << 1$, then [2]

$$x(t) \simeq a \cos(\omega_0 t+\theta_0) \quad , \quad a^2 = 6\alpha/(7\beta^2) \quad , \quad \theta_0 = \text{integration constant} \quad . \tag{14}$$

A considerably more complicated example, arising in epidemiology [6], is

$$Lu = \ddot{u} + (\alpha+\varphi)\dot{u} + \left(\frac{1}{4}\,\omega^2-\Delta+\psi\right)u = -\gamma\beta(t)\cdot(uv+v^2)$$
$$\gamma v = \dot{u} + bu \quad , \quad \beta(t) = \beta_0 + h\cos\omega t \quad , \quad h_0 < h < \beta_0 \quad , \tag{15}$$

where h_0, b, γ, β_0 are fixed positive constants, $\alpha = \alpha(h)$, $\Delta = \Delta(h)$ are known parameters and $\varphi=\varphi(\omega t,h)$, $\psi = \psi(\omega t,h)$ known periodic functions of period $T = 2\pi/\omega$. The 1/2-subharmonic resonance is singly degenerate. If $c = h - h_0 > 0$ is sufficiently small, then

$$u(t) \simeq a \cos\left(\frac{1}{2}\,\omega t+\theta_0\right) \quad , \quad a^2 \simeq c\omega R/(4\gamma^2) \quad , \quad \theta_0 = \text{integration constant} \quad , \tag{16}$$

where $R = R(h_0)$ is in principle a known parameter. Unfortunately a good estimate of R is very difficult to obtain, because R depends strongly on both $\lambda_1 = \lambda_1(c)$ and $\lambda_2 = \lambda_2(c)$ of the generating equation $Lu = 0$, and an efficient constructive algorithm for the determination of λ_1, λ_2 near $c = 0$ is not yet available. In order to illustrate the problems involved, assume that the recurrence (10) is equivalent to (15) at least locally, i.e., assume that the functions f, g in (10) are known in some neighbourhood of the 1/2-subharmonic resonance. Let the general solution of (10) be (cf. [1])

$$x_n = G(n,x_0,y_0,c) \quad , \quad y_n = F(n,x_0,y_0,c) \quad . \tag{17}$$

If the functions F, G are extendable to continuous values of n, then there exist two differential equations (for isolated subharmonic resonances observed in natural evolution processes this property is more a rule than an exception):

$$\frac{dx}{dn} = \bar{g}(x,y,c) \quad , \quad \frac{dy}{dn} = \bar{f}(x,y,c) \quad , \tag{18}$$

whose general solution is also given by (17). In such a case it is in principle possible to transform a non-autonomous differential equation like (13) or (15) into the more easily solvable autonomous system of equations (18). The required transformation is, of course, not only explicitly time-dependent, but it involves also the characteristic exponents $\lambda_1(c)$ and $\lambda_2(c)$.

Conclusion

Subharmonic resonances of a second order dynamic system with periodic coefficients can be characterized by means of the eigenvalues of an equivalent recurrence. There exist several qualitatively distinct types of subharmonic resonances, each requiring a different algorithm for an efficient approximation of the corresponding solutions.

References

1. I. Gumowski, C. Mira: *Dynamique Chaotique* (Cêpadues Editions, Toulouse 1980)
2. I. Gumowski, R. Thibault: *Dynamic Systems with a Singular Parametric Resonance* (Equadif-78, Firenze 1978) pp.91-98
3. Iu. Mitropolskii, I.G. Kozubovskaia: "The problem of nonlinear oscillation in nonlinear systems" (in Russian), Mat. Fiz. *20*, 37-45 (1976)
4. N.N. Bogoliubov, Iu. Mitropolskii: *Asymptotic Methods in the Theory of Nonlinear Oscillations*, 2nd ed. (FIZMATGIZ, Moscow 1958) (in Russian)
5. R. Thibault: "Dynamic systems: Study of a Degenerate Subharmonic Bifurcation in a 2nd-Order Equation by the Separation of Solutions into Periodic and Antiperiodic Parts", Proc. 8th ICNO, Prague (1978) pp.695-700
6. D. Dietz: *The Incidence of Infectious Diseases under the Influence of Seasonal Fluctuations*, in Mathematical Models in Medicine, ed. by J. Berger, J. Bühler, R. Repges, P. Tautu, Springer Lecture Notes in Biomathematics, Vol.11 (Springer, Berlin, Heidelberg, New York 1976) pp.1-15

Chapter 2
Applications in Physics

2.1 Critical Phenomena in Solid-State Physics

On the Bifurcation of Certain Kam Tori in the Standard Mapping

S. Aubry
C.E.N. Saclay, Laboratoire Léon Brillouin, BP 2, F-91190 Gif sur Yvette, France
and
D.R.P., Université Pierre et Marie Curie, F-75231 Paris Cédex 05, France

We study the ground-state $\{u_i\}$ of a one-dimensional chain of atoms with energy

$$\phi(\{u_i\}) = \sum_{i=-\infty}^{+\infty} [\lambda V(u_i)+W(u_{i+1}-u_i)] \tag{1}$$

where u_i is the abscissa of the ith atom. $V(u_i)$ is an analytic periodic and even potential with period 2a, $W(u_{i+1}-u_i)$ is an analytic convex potential which couples neighbouring atoms. A ground-state is a particular solution of the equation

$$\frac{\partial\phi}{\partial u_i} = \lambda V'(u_i) + W'(u_i-u_{i-1}) - W'(u_{i+1}-u_i) = 0 \quad . \tag{2}$$

This equation can be interpreted as the evolution equation of a canonical system with a discrete time i

$$\begin{pmatrix} p_{i+1} \\ u_{i+1} \end{pmatrix} = T\begin{pmatrix} p_i \\ u_i \end{pmatrix} = \begin{pmatrix} p_i+\lambda V'(u_i) \\ u_i+W'^{-1}(p_i+\lambda V'(u_i)) \end{pmatrix} \tag{3}$$

for which ϕ is the action and $p_i = W'(u_i-u_{i-1})$ the conjugate variable of u_i. u_i is mapped modulo 2a so that this mapping T is a diffeomorphism of the cylinder $\mathbb{R}\times{}^1T$. When $W(u)=1/2u^2$ and $V(u)=1/4\pi^2 \cos 2\pi u$, (3) becomes the standard mapping studied by B.V. CHIRIKOV [1] or J.M. GREENE [2]. With fixed boundary condition

$$\lim_{\substack{N\to+\infty \\ N'\to-\infty}} \frac{1}{N-N'} (u_N-u_{N'}) = \ell \quad \text{and} \quad \frac{\ell}{2a} \text{ irrational} \tag{4}$$

it is proved that there exists a hull function $f_\lambda(x)$ which monotonously increases such that the configurations

$$u_i = f_\lambda(i\ell+\alpha) = i\ell + \alpha + g_\lambda(i\ell+\alpha) \tag{5}$$

with arbitrary phase α are the ground states of (1). $g_\lambda(x) = f_\lambda(x) - x$ is then a periodic odd function with the same period as V.

For most irrationals $\ell/2a$, the Kolmogorov-Arnold-Moser theorem allows proving that function f is analytic for λ smaller than a certain $\lambda_2(\ell)$. The trajectories

(u_i, p_i) which represent these ground states are dense on a closed curve on which the transformation T has a rotation number $\ell/2a$. This curve is called a KAM torus.

For λ larger than a certain λ_1 it is proved that a) f has no continuous part, b) f is discontinuous on a dense set of points, c) it has two determinations f_ℓ and f_r which are, respectively, left or right continuous and are equal except at the discontinuities.

This family of functions $f_\lambda(x)$ depends continuously on λ and this proves that the KAM torus for $\lambda < \lambda_2(\ell)$ continuously transforms into a Cantor set for $\lambda > \lambda_1$ which is parametrized as

$$U = f(t)$$
$$P = W'[f(t)-f(t-\ell)] \qquad (6)$$

with the two determinations of f.

This Cantor set is a minimal T invariant closed set.

It is conjectured that the values of λ_1 and $\lambda_2(\ell)$ can be improved in order to be equal to a common $\lambda_c(\ell)$ which is called stochasticity threshold. At $\lambda_c(\ell)$, this KAM torus bifurcates into many invariant closed sets in which the transformation T is generally stochastic. One of these sets is the above defined Cantor set in which the transformation T is not stochastic (for details see [3]).

References

1. B.V. Chirikov: Phys. Reports *52*, 263 (1979)
2. J.M. Greene: J. Math. Phys. *20*, 1183 (1979)
3. S. Aubry: In *Solitons and Condensed Matter Physics*, ed. by A.R. Bishop, T. Schneider, Springer Series in Solid State Sciences, Vol.8 (Springer, Berlin, Heidelberg, New York 1978) p.264
   S. Aubry: "The devil's staircase transformation in incommensurate lattices", in Seminar on the Rieman Problem, Spectral Theory and Complete Integrability, 1978/79, ed. by D.V. Chudnovsky
   S. Aubry, G. André: Annals of the Israel Society *3*, 133 (1980), ed. by L.P. Horwitz and Y. Ne'eman
   S. Aubry: Intrinsic Stochasticity in Plasmas (ed. by G. Laval and D. Grésillon, Edition de Physique 1979) p.63
   S. Aubry: in preparation

MO Stochasticity Criterion

R. Caboz
Université de Pau, Laboratoire de Physique Appliquée, I.U.R.S.
Avenue Philippon, F-64000 Pau, France

A. Lonke
Ben Gurion University
Beer Sheva, Israel

1. Introduction

Using MORI's method [1], MO has studied several hamiltonian systems with a definite and small degree of freedom n .

These systems all have the following hamiltonian :

$$H(q,p) = \sum_{i=1}^{i=2} \frac{1}{2} p_i^2 + \frac{1}{2} q_i^2 \omega_i^2 + \lambda V(q_1, q_2) . \qquad (1)$$

MO gave a criterion which permits us to find (inside the potential well i.e. for $E < E_c$; E_c = escape energy) the threshold E_0 above which a certain amount of stochasticity does appear [2] .

We have shown in a preceding article the nature of the link between MO-MORI's method and the classical moment problem [3] .

MO's criterion which seems to be rather effective was recently analyzed in the scientific literature [4] [5] [6] .

Michael TABOR suspects that the formalism used by MO may not be suitable for detecting the *"chaotic transition"* [4]. Joseph FORD for his part thinks that such a sharp transition is most unprobable [7] .

So, we have entered upon this study in order to clarify the problem.

2. MO's criterion [2]

Given a hamiltonian system (1) MO introduces the inner product $< F, G >$ of two functions F and G of the dynamic variables q and p as follows :

$$< F , G > = \frac{\int_\Gamma d\Gamma \; F^+ \, G \, \delta(H - E)}{\int_\Gamma d\Gamma \; \delta(H - E)} \tag{2}$$

n = degree of freedom number
$d\Gamma = dq_1 \dots dq_n \, dp_1 \dots dp_n$
E = value of the hamiltonian H
$0 \leq E < E_c$
E_c = Escape energy
Γ = phase space
δ = Dirac distribution .

MO's work is concerned with n = 2 and

$$F = d_o = \sum_{i=1}^{i=2} \frac{1}{2} p_i^{\,2} + \frac{1}{2} q_i^{\,2} \quad \text{for } t = t_o \tag{3}$$

$$G = d_{-\tau} = \sum_{i=1}^{i=2} \frac{1}{2} p_i^{\,2} + \frac{1}{2} q_i^{\,2} \quad \text{for } t = -\tau \; . \tag{4}$$

With the Liouville operator :

$$L = i \{ H , \} = i \left(\sum_{i=1}^{i=2} \frac{\partial H}{\partial p_i} \frac{\partial}{\partial q_i} - \frac{\partial H}{\partial q_i} \frac{\partial}{\partial p_i} \right) \tag{5}$$

one obtains :

$$< F , G > = < d_o , e^{-i L \tau} d_o > = D(\tau) \; . \tag{6}$$

With the help of MORI's method [1] MO introduces the LAPLACE TRANSFORM $\hat{D}(Z)$ of $D(\tau)$ and finds :

$$\hat{D}(Z) = \left. \frac{\Delta_o^2}{Z + \frac{\Delta_1^2}{Z + \frac{\Delta_2^2}{Z + \frac{\Delta_3^2}{Z + \dots}}}} \right| \quad \text{for } \operatorname{Re} Z > \sigma \; . \tag{7}$$

We write (7) :

$$\hat{D}(Z) = \frac{\Delta_o^2 |}{| Z} + \frac{\Delta_1^2 |}{| Z} + \frac{\Delta_2^2 |}{| Z} + \frac{\Delta_3^2 |}{| Z} + \dots + \frac{\Delta_n^2 |}{| Z} + \dots \tag{8}$$

with

$$\left.\begin{array}{ll|l} \Delta_o^2 = S_o & & \Delta_2^2 = \frac{S_4}{S_2} - \frac{S_2}{S_o} \\ \Delta_1^2 = S_2/S_o & & \Delta_3^2 = \frac{S_4}{S_2}\,\frac{\frac{S_6}{S_4} - \frac{S_4}{S_2}}{\frac{S_4}{S_2} - \frac{S_2}{S_o}} \end{array}\right\} \qquad (9)$$

and

$$S_m = < d_o , L^m d_o > \qquad (10)$$

MO carries out the numerical perturbative calculation of the moments S_o, S_2, S_4 and S_6 and is thus able to give the approximate expression of

$$\hat{D}_3(Z) = \frac{\Delta_o^2|}{|Z} + \frac{\Delta_1^2|}{|Z} + \frac{\Delta_2^2|}{|Z} + \frac{\Delta_3^2|}{|Z} \qquad (11)$$

The author assumes that $\hat{D}_3(Z)$ gives us a correct approximation to $\hat{D}(Z)$.

Returning to the INVERSE LAPLACE TRANSFORM $D_3(\tau)$ of $\hat{D}_3(Z)$, MO finds for the HENON-HEILES and BARBANIS systems thresholds of $E_o = 0,115$ and $E_o = 0,0101$ respectively. Above this onset $D_3(\tau)$ is diverging exponentially in time [2].

That is what the author calls the stochasticity threshold [2].

3. Critical study of the results

3.1 The expression of $\hat{D}_3(Z)$

Starting with the formula (6) one finds for $D(\tau)$:

$$D(\tau) = < d_o , e^{-iL\tau} d_o > = S_o - \frac{S_2 \tau^2}{2!} + \frac{S_4 \tau^4}{4!} - \frac{S_6 \tau^6}{6!} + \ldots + (-1)^n \frac{S_{2n} \tau^{2n}}{(2n)!} + \ldots \qquad (12)$$

and for its LAPLACE TRANSFORM :

$$\hat{D}(Z) = \frac{S_o}{Z} - \frac{S_2}{Z^3} + \frac{S_4}{Z^5} + \frac{S_6}{Z^7} + \ldots + (-1)^n \frac{S_{2n}}{Z^{2n+1}} + \ldots \qquad (13)$$

It is easy to get the "associated continued fraction" of (13) and one finds by identification [8] [9] :

$$\hat{D}(Z) = \frac{a_o|}{|Z} + \frac{a_1|}{|Z} + \frac{a_2|}{|Z} + \ldots + \frac{a_n|}{|Z} + \ldots \tag{14}$$

with

$$a_o = S_o \quad ; \quad a_1 = \frac{S_2}{S_o} \quad ; \quad a_2 = \frac{S_4}{S_2} - \frac{S_2}{S_o}$$

and

$$a_3 = \frac{S_4}{S_2} \, \frac{\frac{S_6}{S_4} - \frac{S_4}{S_2}}{\frac{S_4}{S_2} - \frac{S_2}{S_o}} \; .$$

The reccurence formulas (15) (16) for the general terms a_{2n} and a_{2n-1} have been given gy GILEWICZ [10] :

$$a_{2n} = \frac{H_n^o \; H_{n-2}^1}{H_{n-1}^o \; H_{n-1}^1} \qquad \text{with } H_{-1}^m = 1 \tag{15}$$

and

$$a_{2n-1} = \frac{H_{n-2}^o \; H_{n-1}^1}{H_{n-1}^o \; H_{n-2}^1} \qquad \text{with } n \geq 1$$

with

$$H_n^m = \begin{vmatrix} y_m & y_{m+1} \ldots & y_{m+n} \\ y_{m+1} & y_{m+2} \ldots & \\ \cdots\cdots & \cdots\cdots & \\ y_{m+n} & y_{m+n+1} & y_{m+2n} \end{vmatrix} \tag{16}$$

and

$$y_m = S_{2m} \tag{17}$$

This expressions (14) (15) (16) and (17) are in fact the same as the ones MORI found [1] but we notice that our *3-step-derivation* is very easy and not time consuming.
We recover the same formula $\hat{D}_3(Z)$ of MORI [1] and MO [2] and it is obvious that all the poles of $\hat{D}_3(Z)$ are purely imaginary, proving that $D_3(\tau)$ can never diverge exponentially in time, which contradicts MO's conclusions.

3.2 Would a better approximation be useful ?

One could take a better approximation $\hat{D}_n(Z)$ (with $n > 3$) of $\hat{D}(Z)$. BREZINSKI [11] was able to prove that all the poles of $\hat{D}_n(Z)$ are purely imaginary for every finite n . The only — but yet not solved — problem is to obtain the exact form of $\hat{D}(Z)$ without making any assumptions i.e. for $n \to \infty$.

3.3 D(τ) is bounded

For $E < E_c$ the functions F and G given by (3) and (4) are bounded and as a result their inner product possesses the same property.

D(τ) being bounded can never diverge exponentially in time : this has been previously conjectured by TABOR [4] . We come to the conclusion that we are now able *to reject definitively MO's criterion* ; but we shall try in a following article to show *how interesting* MO-MORI's method can be for the study of hamiltonian systems with a finite and small degree of freedom.

References

[1] MORI H. Progress in Theoretical Physics (Jap.) 34 (1965) pp.399-416

[2] MO K.C. Physica 57 (1972) pp.445-454

[3] LONKE A. and CABOZ R. Physica 99 (1979) pp. 350-356

[4] TABOR M. Advances in Chemical Physics - preprint - October 1979

[5] TREVE Y.M. Topics in Non-Linear Dynamics
A.I.P. Conference Proceeding n°46 (1978) p.177

[6] CERJAN C. and REINHARDT W.
J. Chem. Phys. 71 (4) 1979 pp.1819-1831

[7] FORD J. Private communication 21.11.1978

[8] BAKER G.A. Jr. Essentials of Padé Approximants
Academic Press N.Y. 1975

[9] GILEWICZ J. Lecture Notes in Mathematics n°667 - 1978 - SPRINGER

[10] GILEWICZ J. Private communication 22.4.1980

[11] BREZINSKI C. Private communications 7.5.1980 and 16.5.1980

Singularities in Saw Numerical Simulations

F.C. Cuozzo

Université de Toulon, Laboratoire d'Automatique et d'Informatique Appliquées, Château Saint-Michel, F-83130 La Garde Cédex, France

E.E. Cambiaggio

Laboratoire d'Electronique, Université de Nice
F-06034 Nice Cédex, France

Introduction

In the numerical simulation techniques presented in (1) and (2), it is necessary to bring particular attention to some singular points, which may be classified according to the nature of the substrate.

A. In the case of an isotropic substrate, there are (fig.1) :

- the points located at the interface substrate-air (with or without mass-loading)
- the points located on corners (90° corner and 270° corner)
- the steps or grooves, of variable size.

B. In the case of anisotropic piezoelectric substrates, there are . (fig.2)

- the interface points (free surface, metallized surface with or without mass-loading)
- the points located at electrical discontinuities (short-circuit edge, one strip or several strips)
- the points located at mechanical discontinuities (etching, mass-loading).

Fig. 1

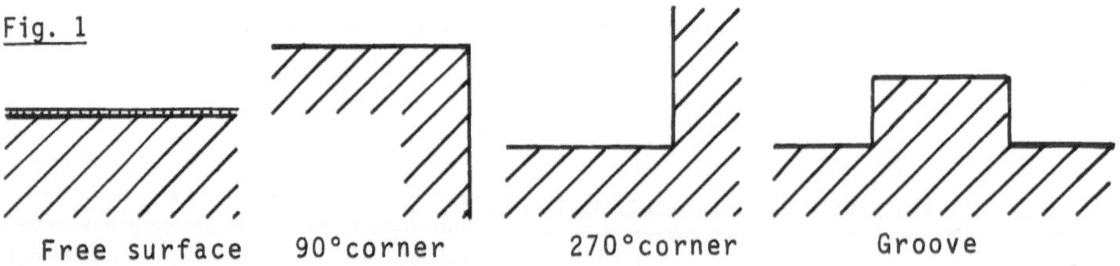

Fig. 2

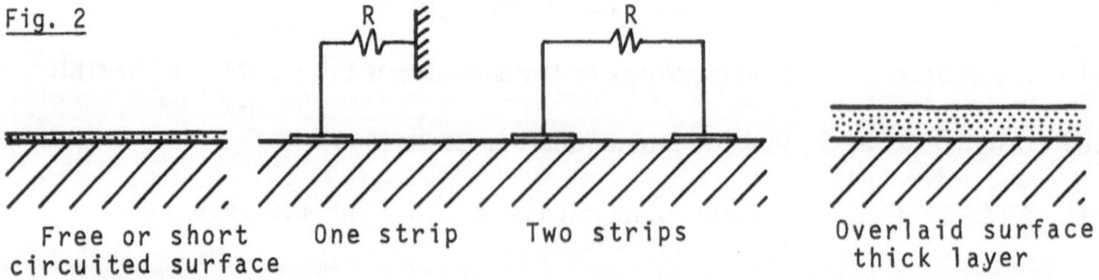

A. Isotropic Substrates

1. Principle

The numerical simulation algorithms on isotropic substrates have been described by many authors(3-8).These algorithms are based upon the finite difference approximations of fundamental equations.

The implementation of a numerical simulation is made as follows.

- The equations of the problem are written, and classified into two categories
  . Dynamic Fundamental equations
  . Equations of boundary conditions.
- By physical considerations, like symmetries, choice of reference axis, isotropy, etc..., the problem is reduced to a problem in the sagittal plane.
- The space and time variables are made descret, and a space grid with square meshes of size h is defined.
- By application of Taylor's formula, partial differential operators are replaced by difference operators.
- The dynamic fundamental equation is hyperbolical, and the problem is an "initial value problem". The finite difference approximation of this equation leads to an explicit formula, by which the mechanical displacement $\vec{u}$ of a node, at a discrete time is a linear combination of displacement of surrounding nodes at the two previous discrete times.
- The evolution of system state is computed recursively, from an initial solution.
- However, the implementation of such a process requires particular care
  i) in the finite difference approximation of boundary conditions
  ii) in the choice of initial conditions
  iii) in the choice of increments, in order to satisfy the stability conditions.

In our previous works, we have been studying many geometries, with various boundary conditions. So, in this paper we discuss only the first point, i.e. the boundary conditions.

Some discussions about ii) and iii) have been presented in other papers. (4,5,9,10)

2. Fundamental Equation

From this fundamental equation, which depends only on the physical properties of the substrate, is built the main part of the algorithm

$$\rho \frac{\partial^2 u_j}{\partial t^2} = C_{ijkl} \frac{\partial^2 u_k}{\partial x_i \partial x_l} \qquad i,j,k,l = 1,2,3 \qquad (1)$$

with
- u_i : component of mechanical displacement along direction x_i
- C_{ijkl} : element of stiffness tensor
- ρ : density
- x_i : space variable
- t : time

3. Boundary Conditions

In the case of isotropic substrate, the boundary conditions may be classified in two categories (with or without mass-loading).

3.1. Surface Without Mass-Loading

3.1.1. Free Surface (Fig.3)

Surface nodes must satisfy both the equation (1) and the boundary condition, i.e. the surface is traction-free

$(T_{3j})_{x_3=0} = 0$ where

$$T_{ij} = C_{ijkl} \frac{\partial u_k}{\partial x_l} \quad \text{is the stress tensor .} \qquad (2)$$

The difficulty is overcome by splitting the surface line into an actual line (J) to which eq.(1) is imposed, and an image line (J-1), to which boundary conditions are imposed.

This yields to the finite difference formula

$$\left.\begin{aligned} u_1(i,j-1,t) &= u_1(i,j+1,t) + u_3(i+1,j,t) - u_1(i-1,j,t) \\ u_3(i,j-1,t) &= u_3(i,j+1,t) + \left[1 - \frac{2v_t^{2*}}{v_l^{2*}}\right] u_1(i+1,j,t) - u_1(i-1,j,t) \end{aligned}\right| \quad (3)$$

where i,j denotes the node position, and t the discretized time.

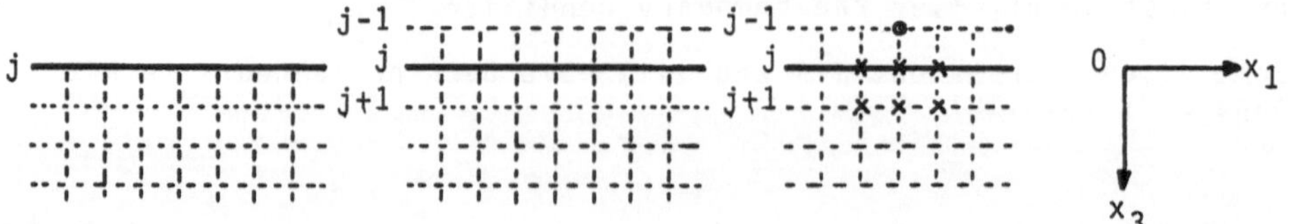

Fig. 3

---

*Defined in Appendix I.

3.1.2. 90^0 and 270^0 Corners (Fig. 4)

The principle of supplementary nodes is used in the case of 90° and 270° corners (4,5,6,7).

For the 90° corner, we get three supplementary nodes A, B and C, related to node P, whereas for the 270° corner, we get only 1 supplementary node, related to node M,N,P.

So, the first case yields to a system with more unknowns than equations, whereas the second case yields to a system with more equations than unknowns.

In both cases, the difficulty is overcome in the same manner : The corner is assumed to have a very little oblique face, and stresses are considered to be null in the direction $\vec{n}$ (normal to AC in the first case, and normal to MN in the second).

This leads for nodes A, C and B to the following equations :

node C :

$$\overline{u}(i+1,j,t) = \tilde{A}_4\overline{u}(i,j,t) - \tilde{A}_5\overline{u}(i-1,j,t) + \tilde{A}_6\left[\overline{u}(i,j,t)-\overline{u}(i,j+1,t)\right] \quad (4)$$

node A :

$$\overline{u}(i,j-1,t) = \tilde{A}_1\overline{u}(i,j,t) - \tilde{A}_2\overline{u}(i,j+1,t) + \tilde{A}_3\left[\overline{u}(i,j,t)-\overline{u}(i-1,j,t)\right] \quad (5)$$

node B :

$$\overline{u}(i+1,j-1,t) = \overline{u}(i,j,t) + \tilde{A}_7\left[(\overline{u}(i,j-1,t)-\overline{u}(i+1,j,t)\right] \quad (6)$$

$\overline{u}$ and $\tilde{A}_i$ are defined in Appendix I.
In the case of 270° corner, this leads to an analogous equation.

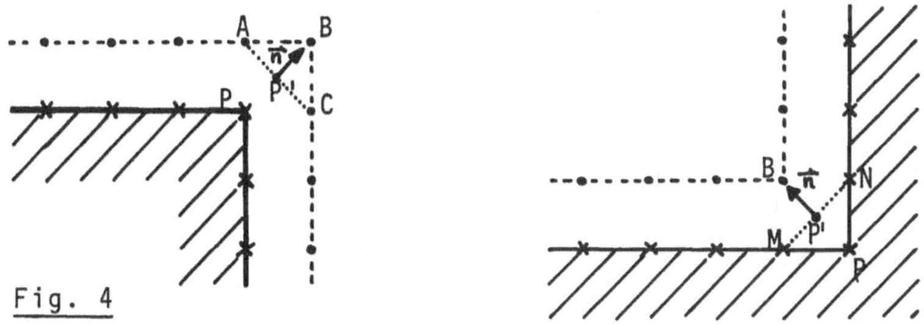

Fig. 4

3.1.3. Steps and Grooves (Fig. 5)

Here, it is necessary to consider two cases

- large steps or large grooves (i.e. $e > 2h$, where h is the mesh size). The step or the groove is considered as an association of elementary configuration presented in the previous paragraph.

Example : upward step = free horizontal surface + 270° corner + free vertical surface + 90° corner + free horizontal surface.

- small steps or grooves (i.e. $e < 2h$) :

The algorithm cannot be used here, and other algorithms must be used (perturbation methods for example (11)).

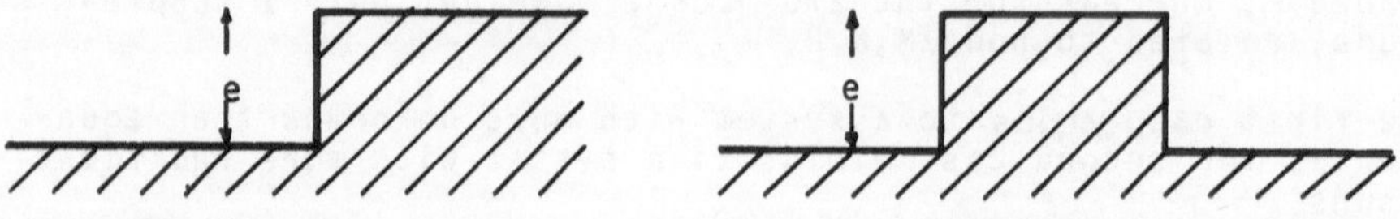

Fig. 5

3.2. Overlaid Surface (Mass-Loading Effects)

3.2.1. Thick Layers ($e > 2h$) (Fig.6)

In order to take into account the layer inerty, it is necessary to sample the layer with two or more lines of nodes (13). In this case, layer nodes must verify both fundamental equations and boundary conditions i-e a continuity of mechanical displacements and stresses. These conditions lead to particular interface formula. It must be noted that the wave which propagates near the interface is different from the wave at a free surface (12).

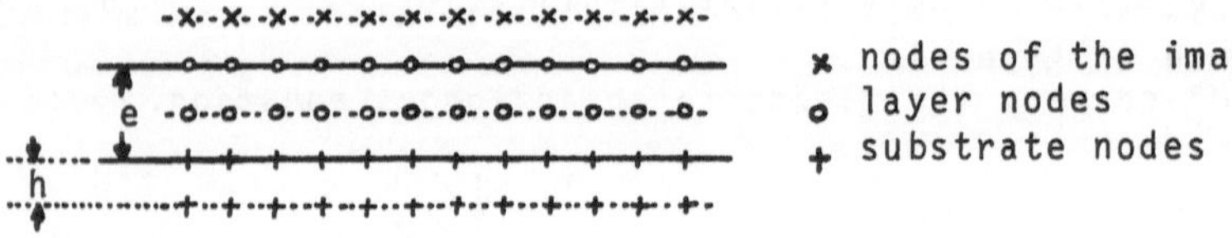

Fig. 6

3.2.2. Thin Layers (Fig. 7)

For the study of thin layers, two ways may be considered :

i) a reduction of mesh size h, with, as a consequence, a considerable increase of computing resources,

ii) a modification of boundary conditions in order to take into account layer intertia.

An approximation of layer inerty had been proposed, (13) into which the surface stresses are

$$T_{3j} = \rho' e \frac{d^2 u_j}{dt^2} \qquad (7)$$

where ρ' is an equivalent density of the layer. In this approximation, the layer is considered as an association of independent elementary masses.

Taking into account the finite width of the layer, this yields to a linear system.

$$
\begin{aligned}
A\,u_1\,(i,j-1,t) + B\left[u_3(i-1,j-1,t) - u_3(i+1,j-1,t)\right] &= \alpha(i,j) \\
C\,u_3\,(i,j-1,t) + B\left[u_1(i-1,j-1,t) - u_1(i+1,j-1,t)\right] &= \beta(i,j)
\end{aligned}
\qquad (8)
$$

where $\alpha(i,j)$ and $\beta(i,j)$ are finite difference expressions obtained from nodes near the surface, and A,B,C are expressions in terms of elastic properties of the substrate.(Appendix II)

Displacements of supplementary nodes are the solution of the system.

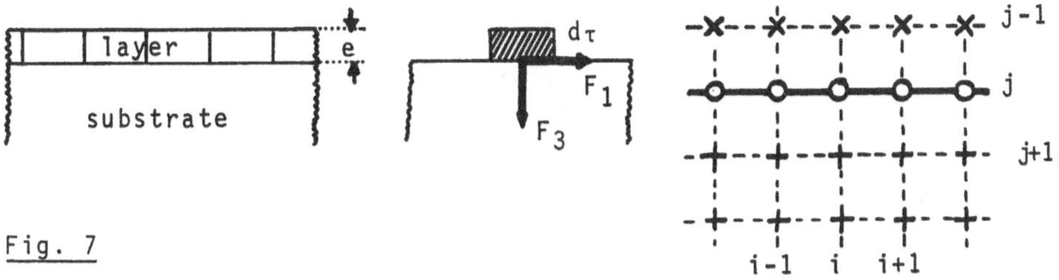

Fig. 7

B. Anisotropic and Piezoelectric Substrates

1. Principle

A numerical simulation algorithm for high coupling piezoelectric crystals has been presented elsewhere [2,14].

Its principle is analogous to the one depicted for an isotropic substrate, but in this case, there is an electrical potential V associated with the mechanical wave.

The equations of the problems are modified by

i) a supplementary fundamental equation for the elasto piezoelectric coupling between $\vec{u}$ and V

ii) electrical boundary conditions (in addition to the previous mechanical conditions).

The coupling equation is an elliptic differential equation. Particular boundary conditions lead to a Dircklet's problem, which is solved by linear iterative method.

The complete algorithm for the simulation of piezoelectric wave propagation is obtained by an association of two distinct process, the first with iterations over time, and the second with iterations over space.

2. Fundamental Equations (15)

$$\rho \frac{\partial^2 u_j}{\partial t^2} = C_{ijkl} \frac{\partial^2 u_k}{\partial x_i \partial x_l} + e_{kij} \frac{\partial^2 V}{\partial x_i \partial x_k} \tag{9}$$

$$0 = e_{ikl} \frac{\partial^2 u_k}{\partial x_i \partial x_l} - \varepsilon_{ik} \frac{\partial^2 V}{\partial x_i \partial x_k} \tag{10}$$

(9) is dynamic fundamental equation for piezoelectric crystal

(10) is elastopiezoelectric coupling equation

where e_{ijk} are elements of piezoelectric tensor

ε_{jk} are elements of dielectric tensor

V is the electrical potential associated to the mechanical displacement $\vec{u}$.

3. Boundary Conditions

Various types of conditions may be imposed to the surface state (free or overlaid substrate, conducting or not conducting layer, thin or thick layer, etc...).

We have studied essentially metallic layers and in this study, we have dissociated its mechanical effects (mass-loading) from its electrical effects (shorting).

So, for the study of electrical effects, the metallic layer is assumed to be massless and perfectly conducting. The mechanical conditions for a mechanically free surface can be used.

3.1. Massless Layer

3.1.1. Homogeneous Surface

This surface may be either entirely free, or entirely short circuited by a massless layer.

- Mechanical interface conditions :

$$T_{3j} = 0 = C_{ijkl} \frac{\partial u_k}{\partial x_l} + e_{3kl} \frac{\partial V}{\partial x_k} \tag{11}$$

- Electrical interface condition

i) free surface

The normal electrical displacement D_3 is continuous across a free interface i-e.

$$D_3 = \hat{D}_3 \tag{12}$$

with

$$D_3 = e_{3kl} \frac{\partial u_k}{\partial x_k} - \varepsilon_{3k} \frac{\partial V_{in}}{\partial x_k} \quad \text{inside the substrate}$$

(13)

$$\hat{D}_3 = -\varepsilon_0 \frac{\partial V_{out}}{\partial x_3} \qquad \text{outside the substrate ,}$$

ii) <u>short circuited surface</u>

$$V_{in} = V_{out} = 0 \quad \text{for } x_3 = 0 \, . \tag{14}$$

These equations may be regrouped :

$$(D)\ \underline{u}\ (i,j-1,t) = \begin{bmatrix}\alpha\\ \beta\\ \gamma\end{bmatrix} \quad \text{with } \underline{u} = \begin{vmatrix}u_1\\ u_3\\ V\end{vmatrix} \tag{15}$$

where (D), α, β and γ are defined in Appendix III.

Case i) and ii) may be applicable simultaneously to different regions of the surface, in the case of a partly coated surface.

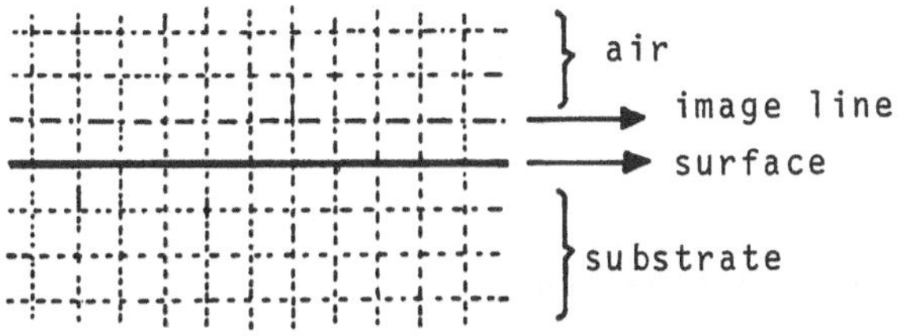

Fig. 8

3.1.2. Surface with Local Metallizations (Metallic Strips)

In addition to local conditions presented in paragraph 3.1.1. it is necessary to take into account the particular conditions imposed to the strip.

For a single strip (fig.9-a), three cases must be considered according to the resistor value

$R = 0 \Rightarrow V_S = 0$ The strip potential is always equal to zero (grounded strip)

$R = \infty \Rightarrow Q = 0$ The strip is insulated, its total electrical charge Q is always equal to zero, but its potentials V_S varies as the waveforms under it.

The application of Gauss's theorem around the strip (fig.9-b) yields to a finite difference formula, in which V_S is a function of potentials and displacements of surrounding nodes.

$R \neq 0$ In this case, $V_S \neq 0$, $Q \neq 0$.

A finite difference approximation of ohm's law

$V_S = -R \frac{dQ}{dt}$ yields to a recursive determination of Q, which allows the determination of V_S by the previous approximation of Gauss's theorem.

For a set of two strips, interconnected via a resistor, we get the supplementary equations

$$V_{S1} \neq V_{S2} \neq 0$$
$$Q_1 \neq Q_2 \neq 0$$

$$V_{S1} - V_{S2} = R\,\frac{dQ_1}{dt}\,.$$

Simulations have been carried out for a variable number of strips, n = 1,2,4 and 8.

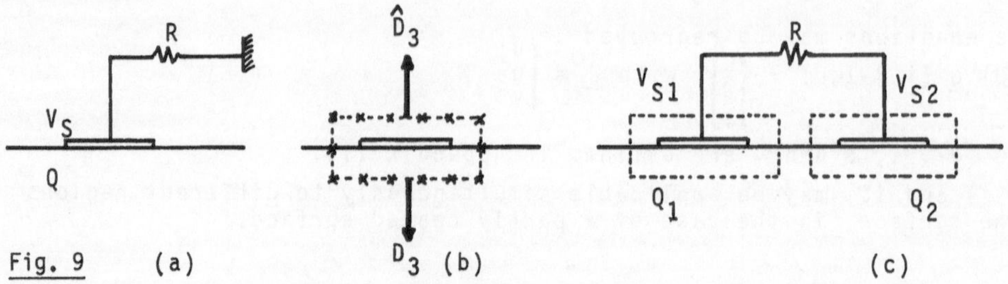

Fig. 9 (a) (b) (c)

3.2. Weighty Strips (Mass-Loading)

When the mass-loading effect of a thin layer is non-negligible, the study presented in the previous paragraph is available, but the mechanical interface condition $T_{3j}=0$ must be replaced by a condition analogous to the one presented in § A.3.2.2.

The algorithm has not been developped in details, but one can distinguish the cases of conducting layer or insulating layer.

For an insulating layer, the piezoelectric substrate may be replaced by an equivalent isotropic substrate (15,16). This allows a reduction of computing resources, and algorithms of § A.3.2.2. are available.

For a conducting strip, it is necessary to develop a complete algorithm, taking into account both electrical boundary conditions and mass-loading. This algorithm may be developed from an association of methods presented in previous paragraphs.

Conclusion

A survey of the various singularities encountered in numerical simulation of acoustic surface wave propagation has been presented. In every case, difficulties are printed out, and, when the problems have been solved solutions has been presented.

Appendix 1

$$\tilde{A}_1 = \begin{bmatrix} 4/3 & 0 \\ 0 & 4/a \end{bmatrix} \quad \tilde{A}_2 = \begin{bmatrix} 1/3 & 0 \\ 0 & y^4/a \end{bmatrix} \quad \tilde{A}_3 = \begin{bmatrix} 0 & 2/3r \\ 2y^2/ar & 0 \end{bmatrix}$$

$$\tilde{A}_4 = \begin{bmatrix} 4/a & 0 \\ 0 & 4/3 \end{bmatrix} \quad \tilde{A}_5 = \begin{bmatrix} y^4/a & 0 \\ 0 & 1/3 \end{bmatrix} \quad \tilde{A}_6 = \begin{bmatrix} 0 & 2y^2r \\ 2r/3 & 0 \end{bmatrix}$$

with
$$r = \frac{h}{l}\ ,\ y = (1 - \frac{2v_t^2}{v_l^2}) \quad \text{and} \quad a = 4 - y^4$$

$$\tilde{A}_7 = \begin{bmatrix} A_1 & A_2 \\ -A_2 & A_3 \end{bmatrix} \quad \text{with} \quad A_1 = 1 - \frac{2r^2}{1-2r\ (\frac{\lambda+\mu}{\mu})+r^2}$$

$$A_2 = \frac{-2r}{1-2r\ (\frac{\lambda+\mu}{\mu}) + r^2} \qquad A_3 = \frac{2}{1-2r\ (\frac{\lambda+\mu}{\mu}) - r^2} - 1 \quad ,$$

l is the time increment, λ and μ are Lame's constants.
V_l and V_t are respectively longitudinal and transversal Rayleigh wave velocities
$$V_l^2 = \frac{\mu}{\rho}\ ,\quad V_t^2 = \frac{\lambda+2\mu}{\rho}\ .$$

Appendix II

$$A = V_t^2\ (\frac{1}{h} + \frac{\rho}{2\rho' e}) \qquad B = \frac{V_l^2 - V_t^2}{h} \qquad C = V_l^2\ (\frac{1}{h} + \frac{\rho}{2\rho' e})$$

$$\begin{aligned} \alpha(i,j) = {} & \frac{\rho V_t^2}{2\rho' e}\Big[u_3(i+1,j,t) - u_3(i-1,j,t) + u_1(i,j+1,t)\Big] \\ & + \frac{V_t^2}{h}\Big[2u_1(i,j,t) - u_1(i,j+1,t)\Big] \\ & + B\Big[u_3(i-1,j+1,t) - u_3(i+1,j+1,t)\Big] \\ & - \frac{V_l^2}{h}\Big[u_1(i+1,j,t) + u_1(i-1,j,t) - 2\,u_1(i,j,t)\Big] \end{aligned}$$

$$\begin{aligned} \beta(i,j) = {} & \frac{\rho}{2\rho' e}\Big[(V_l^2-2V_t^2)\ u_1(i+1,j,t)-u_1(i-1,j,t)+V_l^2 u_3(i,j+1,t)\Big] \\ & + \frac{V_t^2}{h}\Big[u_3(i+1,jt)+u_3(i-1,j,t)-2u_3(i,j,t)\Big] \\ & + \frac{V_l^2}{h}\Big[2u_1(i,j,t)-u_3(i,j+1,t)\Big]+B\Big[u_1(i-1,j+1,t)-u_1(i+1,j+1,t)\Big] \end{aligned}$$

Appendix III

$$\begin{aligned} \alpha = {} & C_{35}\Big[u_1(i,j+1,t)+u_3(i+1,j,t)-u_3(i-1,j,t)\Big] \\ & + C_{35}u_3(i,j+1,t)+e_{35}V(i,j+1,t) \end{aligned}$$

$$\begin{aligned} \beta = {} & C_{13}\Big[u_1(i+1,j,t)-u_1(i-1,j,t)\Big]+C_{35}u_3(i,j+1,t) \\ & +C_{35}\Big[u_1(i,j+1,t)+u_3(i+1,j,t)-u_3(i-1,j,t)\Big] \end{aligned}$$

$$+e_{13}\left[V(i+1,j,t)-V(i-1,j,t)\right]+e_{13}\left[V(i,j+1,t)\right]$$

γ and matrix D are dependent on assumed conditions.

i) Free surface

$$(D) = \begin{bmatrix} C_{35} & C_{35} & e_{35} \\ C_{35} & C_{33} & e_{35} \\ e_{35} & e_{33} & \varepsilon_{33} \end{bmatrix}$$

$$\gamma = e_{35}\left[u_1(i,j+1,t)+u_3(i+1,j,t)-u_3(i-1,j,t)\right]+e_{33}\left[u_3(i,j+1,t)\right]$$
$$-\varepsilon_{33}V(i,j+1,t)-\varepsilon_0\left[V(i+1,j,t)+V(i-1,j,t)+2V(i,j-1,t)-4V(i,j,t)\right]$$

ii) Short circuited surface

$$(D) = \begin{bmatrix} C_{35} & C_{35} & e_{35} \\ C_{35} & C_{33} & e_{33} \\ 0 & 0 & 1 \end{bmatrix}$$

$$\gamma = 2V\,(i,j,t) - V\,(i,j+1,t)$$

References

1 E. CAMBIAGGIO, F. CUOZZO, and E. RIVIER, in "Proceedings of the Fifth Colloquium on Microwave Communication," Akadémiai Kiado, Budapest, June 1973, ET-11.
2 E. CAMGIAGGIO and F. CUOZZO, J. Comp. Phys., 33 n°2, (1979).
3 Z. ALTERMAN and F.C. KARAL, Bull. Seism. Soc. Amer., 58 n° 1 (1968)
4 Z. ALTERMAN and A. ROTENBERG, Bull. Seismol. Soc. Amer. 59, n° 1 (1969).
5 Z. ALTERMAN and D. LOEWENTHAL, Geophys. J. Roy. Soc. 20 (1970),101.
6 M. MUNASINGHE and G.W. FARNELL, in "1972 Ultrasonics Symposium," p. 267, IEEE Cat. 72 CHO 708-8SU,1972.
7 M. MUNASINGHE and G.W. FARNELL, J. Appl. Phys. 44, n° 5 (1973).
8 M. MUNASINGHE and G.W. FARNELL, J. Geophys. Res. 78, n° 14 (1973), 2454.
9 A. ILAN and D. LOEWENTHAL, Geophys. Prospect. XXIV, n° 3 (1976),431
10 A. ILAN, J. Comp. Phys. 29, (1978).
11 B.A. AULD, "Acoustic Fields and Waves in Solids," Wiley, New York/ London, 1973
12 H.F. TIERSTEN, J. Appl. Phys. 20, n° 2 (1969), 770.
13 E. CAMBIAGGIO, Thèse Nice (1978).
14 E. CAMBIAGGIO and F. CUOZZO, in "1975 Ultrasonics Symposium Proceedings," p.444, IEEE Cat. n° 75 CH0994-4SU, 1975.
15 G.W. FARNELL, in "Physical Acoustics" (W.P. MASON and R.N. THURSTON, Eds.), Vol.6, Academic Press, New York/London, 1970.
16 R.C.M. LI, Ultrasonics Symp. Proc. IEEE Cat. 72, CH0708-8SU,(1972)

Monte Carlo Measurement of the Single Vortex Free Energy in the Kosterlitz-Thouless Theory

H.J. Hilhorst, H.N.J. Vogelij, C. van Leeuwen, and B.P.Th. Veltman

Laboratorium voor Technische Natuurkunde, Postbus 5046
2600 GA Delft, The Netherlands

1. Introduction

In this communication we elaborate and apply the equivalences that are known to link the Solid-on-Solid (SOS) model to the two-dimensional XY model and the Coulomb lattice gas. We show that on periodic lattices a simple quantity σ can be defined for the SOS model - playing the role of an interface smoothness - which in the XY model or the Coulomb gas represents the usual renormalized *free vortex fugacity* y for a single free vortex. Monte Carlo measurements of this quantity were carried out on SOS lattices of up to 4096 sites. A rapid rise in y is observed as the temperature passes through its roughening value.

2. SOS Model, XY Model, and Coulomb Gas

2.1 The *SOS model* [1], which describes the behavior of a crystal-fluid interface, has recently gained wide interest in the statistical mechanics of phase transitions, as it was discovered to be closely connected to such intriguing systems as the Coulomb lattice gas [2] and the XY model [3,4]. A general column model on a square $\sqrt{N} \times \sqrt{N}$ lattice is represented by a set of N integer-valued column variables h_i. The upper faces of the columns represent the crystal-fluid interface. With each configuration $h = \{h_i\}$ an energy is associated given by the Hamiltonian

$$H_\ell(h) = \frac{J}{\ell} \sum_{<i,j>} |h_i - h_j|^\ell \; . \tag{1}$$

Here J is an interaction strength and the sum is on all nearest neighbor pairs $<i,j>$. The SOS model corresponds to the case $\ell = 1$. For $\ell = 2$ we obtain the closely related *discrete Gaussian* model, which will also play a role in our considerations. We tacitly assume in the following discussion that the qualitative properties of both models are the same. Throughout, we consider lattices with periodic boundary conditions.

The partition function of the Hamiltonian (1) is defined by

$$Z_\ell = \sum_h{}' \, e^{-\beta H_\ell(h)} \tag{2}$$

where $\beta = 1/k_B T$ with T the column model's temperature. The prime on the summation sign indicates that whenever two interfacial configurations h and h' differ only by a vertical translation, just one of them enters into the sum. The interface is expected to be flat at low T and rough at high T. The transition from flat to rough takes place at a well-defined roughening temperature T_R, as was first argued by BURTON, CABRERA, and FRANK [1].

2.2 It was shown by KNOPS [3] and by JOSE, KADANOFF, KIRKPATRICK, and NELSON [4] that the column models (1) can be mapped onto an *XY model* of the type

$$H_{XY}(\varphi) = \sum_{\langle r,s\rangle} V(\varphi_r - \varphi_s) \; . \tag{3}$$

Here r, s, ... are sites of the dual lattice, and φ_r, φ_s, ... are angle variables ranging between $-\pi$ and π. The function V is a 2π-periodic interaction related to the interaction (1) by

$$e^{-\tilde{\beta}V(\varphi)} = \sum_{n=-\infty}^{\infty} e^{in\varphi-\beta J\ell^{-1}|n|^{\ell}} \tag{4}$$

where $\tilde{\beta} = 1/k_B\tilde{T}$ with $\tilde{T}$ the XY temperature.

2.3 The *Coulomb lattice gas* is defined by the Hamiltonian

$$H_{CG}(k) = -J_{CG} \sum_{\langle i,j\rangle} k_i k_j U(|r_i - r_j|) \tag{5}$$

where J_{CG} is an interaction constant, k is a set of integer-valued charges k_i at the lattice points $\vec{r}_i$, and U(r) is the two-dimensional Coulomb potential, which increases logarithmically with r. CHUI and WEEKS [2] showed that the discrete Gaussian model can be mapped exactly onto a neutral Coulomb gas with $J_{CG} = 2\pi/J$.

3. Vortex Fugacity y and Interface Smoothness $\langle\sigma\rangle$

The breakthrough in the study of the above models is due to KOSTERLITZ and THOULESS (KT) [5,6], who recognized the importance of topological excitations in the XY model. When the XY temperature $\tilde{T}$ is low, these so-called vortices (corresponding to the charges in the Coulomb gas) occur only in bound pairs of opposite sign. However, above a critical temperature $\tilde{T}_c$ the first pairs dissociate.

The KT theory contains an explicitly introduced small parameter $y^2(0)$, which is the bare fugacity of a closely bound pair of vortices. By successive renormalization group iterations one can calculate the effective temperature dependent fugacity $y^2(r;\tilde{T})$ for a pair of vortices a distance r apart. The value of the fully renormalized fugacity $y(\infty;\tilde{T})$ determines whether pair dissociation is possible. For $\tilde{T} \leq \tilde{T}_c$, $y(r;\tilde{T})$ iterates to zero, whereas above $\tilde{T}_c$ it eventually iterates away from zero. It is this quantity y that we consider in this conference contribution.

In the Hamiltonians (1), (3), and (5) the parameter y remains implicit (the vortices have no obvious interpretation in the SOS model). Nevertheless, the properties of the vortex system can be determined. This was done *e.g.* in a Monte Carlo study of the XY model by CHESTER and TOBOCHNIK [8]. In order to make certain XY quantities more easily accessible to computer experiment, several authors have proposed to utilize the above model equivalences. *E.g.* SWENDSEN [9], and VAN DER EERDEN and KNOPS [10], transcribe XY correlation functions to step free energies in the SOS model. Here we present another such transcription, but of a somewhat more subtle kind: it involves the boundary conditions in an essential way and requires that we carefully reconsider the equivalence transformations for finite lattices. Since, to our knowledge, only infinite lattices have been treated in the literature, we briefly indicate the main points of difference in the last section.

Motivated by Monte Carlo results of C. VAN LEEUWEN [7] we introduce for the column models (1) a quantity σ, invariant under vertical translations, and defined as

$$\sigma = \exp 2\pi i N^{-1} \sum_j h_j \qquad (6)$$

whose average is a measure of the interface smoothness. It is obviously very easy to measure σ in any MC simulation. In Sec. 4 we show that translation of the average $\langle\sigma\rangle$ into Coulomb gas language yields

$$\langle\sigma\rangle = \frac{Z'_{CG}}{Z_{CG}} \quad . \qquad (7)$$

Here Z_{CG} is the Coulomb gas partition function, and Z'_{CG} is a similar partition function, except that it refers to a system with one surplus negative charge; hence $\langle\sigma\rangle$ is the free energy of a single free charge.

Upon translating $\langle\sigma\rangle$ to the XY model we expect to obtain the free energy of a single free vortex. Indeed we find

$$\langle\sigma\rangle = \frac{\bar{Z}'_{XY}}{\bar{Z}_{XY}} \qquad (8)$$

where Z_{XY} is the XY partition function, and Z'_{XY} is a similar partition function, except for a system with one surplus vortex of given sign. If at first sight the presence of a single free vortex may seem geometrically forbidden on a torus, in Sec. 4 we show how the particular transformation properties of σ on a finite lattice lead precisely to this effect. The bars in (8) indicate an extra average, *viz.* on "spin wave" boundary conditions, also explained in Sec. 4.

At low temperature T, the average value $\langle\sigma\rangle$ is most easily obtained directly in the SOS model. If $N^{-1}\Sigma_i h_i$ is of order $1/\sqrt{N}$, we have

$$\langle\sigma\rangle = 1 + \mathcal{O}(N^{-1}) \qquad T < T_R \quad . \qquad (9)$$

At high SOS temperature T, the XY temperature $\tilde{T}$ is low and $\langle\sigma\rangle$ can be calculated conveniently from expression (8). Using the KT argument [5,6,11], we have for the single vortex free energy

$$\begin{aligned} \langle\sigma\rangle &\sim N^{1-\frac{1}{2}\pi K(\infty;T)} && T > T_R \\ &\sim 1/\log N && T = T_R \quad . \end{aligned} \qquad (10)$$

Here $K(\infty;T)$ is the renormalized coupling constant of the XY model; it increases from $2/\pi$ to ∞ as T goes from T_R to ∞. Expression (10) may also be obtained from the Coulomb gas representation (7). A factor $N^{-\frac{1}{2}\pi K}$ (with $K = 2/\beta J$) then arises from the "dissociation energy" between the extra negative charge and its positive counterpart, eliminated from the system. From (9) and (10) we see that, depending on the temperature, $\langle\sigma\rangle$ tends to 0 or to 1 with the size of the system.

We have performed a set of exploratory MC measurements of $\langle\sigma\rangle$ for an SOS model on a 64×64 lattice. A specially constructed hardware model (described elsewhere [12]) was used to generate the configurations. For each value of the temperature, after an equilibration run of 10240 trials per site, MC runs were carried out of about 400,000 trials per site (for $\beta J > 0.825$) or 800,000 trials per site (for $\beta J \leq 0.825$). (The efficiency of the particular set of transition probabilities built into the model is about 7% for the temperatures of interest).

We calculated σ for configurations sampled at intervals of 2048 trials per site. The average values so obtained are shown in Fig. 1. Standard deviations have been indicated for $\beta J = 0.775$ and $\beta J = 0.800$, where exceptionally long measurements were taken (8,000,000 and 2,000,000 trials per site, respectively).

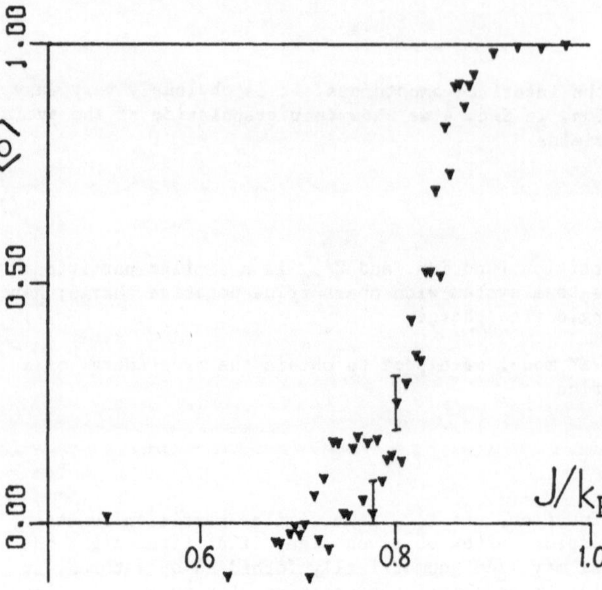

Fig. 1 The surface smoothness <σ> as a function of the inverse temperature J/k_BT from Monte Carlo measurements in a 64×64 SOS model. The steep rise in <σ> near the roughening temperature ($J/k_BT \approx 0.8$) is direct evidence for the occurrence of free vortices in the corresponding XY model (see text).

We see that in the region $\beta J \approx 0.8$, the smoothness <σ> begins to increase sharply from values fluctuating around 0 to values near 1. This observed behavior agrees well with equations (9) and (10). From measurements on a 32×32 lattice we found that <σ> increases with N when $\beta J \gtrsim 0.825$, and decreases below this value. The KT recursion relations do not exclude that a decreasing $y(\sqrt{N};\tilde{T})$ may, at still larger N, increase again. We conclude, therefore, that $J/k_BT_R \lesssim 0.825$ for the SOS model. This bound on T_R agrees with MC work by SHUGARD, WEEKS, and GILMER [13], based on a study of the correlation function $\langle(h_i - h_j)^2\rangle$.

During the MC test the autocorrelation time of σ was measured at values of βJ a distance 0.050 apart. A peak was observed at βJ = 0.800, where relaxation was about 1½ and three times slower than at the neighbouring values βJ = 0.750 and 0.850, respectively (*viz.*, about 10,000 trials per site). More detailed static and dynamic measurements are in progress.

4. Transformation of <σ> on Periodic Lattices

We indicate here the refinements needed to carry out the equivalence transformations mentioned in Sec. 3 on finite periodic lattices, and apply these results to the transformation of <σ>.

4.1 Discrete Gaussian Model to Coulomb Gas

The transformation between these two models is based on finding a new representation for the summations on the column variables [2]. Thereby a new set of discrete variables k_i is introduced, which play the role of the Coulomb charges.

In our case we proceed as follows. To calculate $\langle\sigma\rangle$ we perform Σ'_h in (2) by imposing $h_1 = 0$ and letting all other h_i vary freely through the integers. This approach is equivalent to the localizing field method [2]. The summations on the h_i may be replaced with integrations on continuous variables ν_i if one inserts the appropriate δ-functions. Doing so we obtain

$$\Sigma'_h e^{2\pi i m N^{-1}\Sigma_j h_j} e^{-\beta H_\ell(h)} = \int_{-\infty}^{\infty} \prod_i d\nu_i \int_{-\frac{1}{2}}^{\frac{1}{2}} d\kappa_1 \sum_k e^{2\pi i m N^{-1}\Sigma_j \nu_j} \times e^{2\pi i \Sigma_j k_j \nu_j + 2\pi i \kappa_1 \nu_1} e^{-\beta H_\ell(\nu)} . \tag{11}$$

The newly introduced parameter m equals 0 or 1 in our case of interest. The integration on κ_1 is a consequence of the special role played by h_1. In (11) we pass to the Fourier variables

$$k_{\vec{q}} = N^{-\frac{1}{2}} \sum_j e^{-i\vec{q}\cdot\vec{r}_j} k_j \tag{12}$$

and analogously defined $\nu_{\vec{q}}$. For the discrete Gaussian model ($\ell = 2$) the $\nu_{\vec{q}}$ can then be integrated out. The integral on ν_0 gives a factor $\sqrt{N}\delta(\Sigma_i k_i + \kappa_1 + m)$. Subsequent integration on κ_1 converts the RHS of (11) into

$$Z_{sw} N^{\frac{1}{2}-\frac{1}{2}m^2\pi K} \sum_k \delta(\Sigma_i k_i + m) e^{2\pi K \Sigma_{i<j} k_i k_j U(|\vec{r}_i - \vec{r}_j|)} \tag{13}$$

where $K = 2/\beta J$, Z_{sw} is a nonsingular spin wave partition function, and $U(r)$ is the (approximate) Coulomb potential given explicitly *e.g.* in [2]. The δ-function in (13) imposes that the integer "charges" k_i add up to a total net charge $-m$. Expression (7) for $\langle\sigma\rangle$ is obtained as the ratio of (13) for m = 1 and m = 0.

4.2 SOS to XY Model

The equivalence transformation in this case is a dual transformation. Following [3] and [4] we introduce the 2N variables $n_{ij} = -n_{ji} = h_i - h_j$ for each pair $\langle i,j\rangle$ (ordered according to an arbitrary convention). As we keep $h_1 = 0$ fixed, the system contains only N - 1 degrees of freedom. The N + 1 restrictions to be imposed upon the n_{ij} are

$$\Sigma^{(r)} n_{ij} = 0 \qquad r \text{ a plaquette} \tag{14a}$$

$$\Sigma^{(\alpha)} n_{ij} = 0 \qquad \alpha = x,y . \tag{14b}$$

In (14a) the sum is on the four n_{ij} surrounding a plaquette r. There are N - 1 independent plaqettes. Two more restrictions are furnished by (14b), where the sum is on $\sqrt{N}$ variables n_{ij} forming a closed circle in the α direction around the torus. We now have the summation identity

$$\Sigma' = \int_{-\pi}^{\pi} \frac{d\psi_x}{2\pi} \frac{d\psi_y}{2\pi} \int_{-\pi}^{\pi} \prod_{r=1}^{N} \frac{d\varphi_r}{2\pi} \sum_n e^{i[\Sigma_r \varphi_r \Sigma^{(r)} n_{ij} - \Sigma_\alpha \psi_\alpha \Sigma^{(\alpha)} n_{ij}]} \tag{15}$$

(sum over h on the left-hand side)

into which an extra δ-function has been introduced for the *Nth* plaquette. If this identity is applied to a function of the h_i, and subsequently the n_{ij} in the RHS of (15) are summed out, the original SOS summation is converted into a summation on the XY angles ψ_r and the two special variables ψ_x and ψ_y.

As a stepping stone towards our result we define two generalized Hamiltonians,

$$H_1(h;\vec{n}) = J \sum_{<i,j>} |h_i - h_j + \vec{\vartheta}_{ij}\cdot\vec{n}| , \tag{16a}$$

$$H_{XY}(\varphi;\vec{\psi}) = \sum_{<r,s>} V(\varphi_r - \varphi_s + \vec{\vartheta}'_{rs}\cdot\vec{\psi}) . \tag{16b}$$

Here $\vec{\vartheta}_{ij}\cdot\vec{n} = \vartheta^x_{ij} n_x + \vartheta^y_{ij} n_y$; ϑ^α_{ij} equals unity if $<i,j>$ crosses a fixed circle on the dual lattice parallel to the α axis, and is zero otherwise; and the n_α are arbitrary integers. Analogous definitions hold for $\vec{\vartheta}'_{rs}\cdot\vec{\psi}$, except that the ψ_α are the angles encountered above. $H_1(h;\vec{n})$ is a "step" Hamiltonian with step height n_α in the α direction. $H_{XY}(\varphi;\vec{\psi})$ describes an XY system containing an extra spin wave with wave-vector $\vec{\psi}/\sqrt{N}$, incommensurate with the periodic boundary conditions.

It is interesting, first of all, to apply the sum representation (15) to the partition function of $H_1(h;\vec{n})$. Summing out the n_{ij} and doing some extra algebra we obtain the following curious identity between partition functions, in obvious notation,

$$Z_1(\vec{n}) = \frac{1}{(2\pi)^2} \int_{-\pi}^{\pi} d\vec{\psi}\, e^{i\vec{n}\cdot\vec{\psi}} Z_{XY}(\vec{\psi}) . \tag{17}$$

We now apply the same procedure to the average of σ with respect to $H_1(h;0)$. To this end we express the factor $\exp(2\pi i N^{-1} \Sigma_j h_j)$ in the n_{ij} by writing $h_k = \Sigma^{(P_k)} n_{ij}$, where the sum is on a path P_k linking site k to the origin. The result for $<\sigma>$ is expression (8), where the bar indicates the average $(2\pi)^{-2} \int_{-\pi}^{\pi} d\vec{\psi}$, and where Z'_{XY} is the partition function of the Hamiltonian

$$H'_{XY}(h;\vec{\psi}) = \sum_{<r,s>} V(\varphi_r - \varphi_s + \chi_{rs} + \vec{\vartheta}'_{rs}\cdot\vec{\psi}) . \tag{18}$$

Here χ_{rs} equals $-2\pi/N$ times the number of paths P_k that cross the bond $<r,s>$. We are free to choose the path P_k such as to follow as closely as possible the straight line joining site k to the origin. Let R and Φ be the polar coordinates of the bond $<r,s>$. One then easily verifies that, for large R,

$$\begin{aligned} \chi_{rs} &\simeq \frac{1}{R}\sin\Phi \qquad \text{if } <r,s> \text{ is horizontal} \\ &\simeq \frac{1}{R}\cos\Phi \qquad \text{if } <r,s> \text{ is vertical .} \end{aligned} \tag{19}$$

A ground state of the Hamiltonian (18) is obtained, for $\vec{\psi} = 0$, by parallel alignment of all spins φ_r. However, due to the particular form of χ_{rs}, the ground state energy contains the extra contribution of a single vortex at the origin. This is best seen by Taylor expanding (18) for small χ_{rs}. Obviously the ground state is N-fold degenerate: the others are obtained if the φ_r are placed in a vortex pair configuration of which one member annihilates the effect of the built-in vortex at the origin. We expect that these considerations are not basically modified by the presence of an additional soft spin wave $\vec{\psi}/\sqrt{N}$. Finally we emphasize the essential role of the periodic boundary conditions: on an infinite lattice the angle differences χ_{rs} in the Hamiltonian (18) could simply be transformed away by a redefinition of the angles φ_r.

References

1. W.K. Burton, N. Cabrera, and F.C. Frank, Phil.Trans.Roy.Soc. A243, 299 (1951)
2. S.T. Chui and J.D. Weeks, Phys.Rev. B 14, 4978 (1976)
3. H.J.F. Knops, Phys.Rev.Lett. 39, 766 (1977)
4. J.V. José, L.P. Kadanoff, S. Kirkpatrick, and D.R. Nelson, Phys.Rev. B 16, 1217 (1977)
5. J.M. Kosterlitz and D.J. Thouless, J.Phys. C 6, 1181 (1973)
6. J.M. Kosterlitz, J.Phys. C 7, 1046 (1974)
7. C. van Leeuwen, thesis, Delft University of Technology (1977)
8. J. Tobochnik and V. Chester, Phys.Rev. B 20, 3761 (1979)
9. R.H. Swendsen, Phys.Rev. B 17, 3710 (1978)
10. J.P. van der Eerden and H.J.F. Knops, Phys.Lett. A 66, 334 (1978)
11. A.P. Young, J.Phys. C 11, L453 (1978)
12. J.P. van der Eerden, C. van Leeuwen, P. Bennema, W.L. van der Kruk, and B.P.Th. Veltman, J. Appl. Phys. 48, 2124 (1977)
13. W.J. Shugard, J.D. Weeks, and G.H. Gilmer, Phys.Rev.Lett. 41, 1399, 1577 (1978)

Algebraic Method for the Computation of the Partition Functions of Spin Glasses and Numerical Study of the Distributions of Zeros

B. Lacolle

Laboratoire IMAG, B.P. 53X
F-38041 Grenoble Cêdex, France

Abstract

In this paper we summarize some research on the determination of a phase transition in a two-dimensional spin glass. Professor R. Maynard (Grenoble) suggested the original idea of using the partition functions of small systems and the Yang and Lee theorem. We present two distinct parts. The first one includes experimental results on 6 x 6 square lattices and a new determination of the critical temperature as a function of the concentration of negative interactions. For this purpose we assume that the zeros of partition functions are asymptotically distributed on regular curves and we approximate these curves with the numerical values corresponding to the zeros of 6 x 6 small systems. Some analogous experiments have been realized by M. Suzuki [9,10] but only for the Ising Model. The 6 x 6 size was a computing time limitation when we used enumeration methods to calculate partition functions. In the second part we present, with the Spinor Analysis formalism, an algebraic method for the computation of partition functions, which permits us to reduce the amount of computing time in a large proportion.

Part I Experimental Results About a Spin Glass Model

1.1 The Frustration Model of Spin Glass [11]

Our physical model is a plane square lattice with n rows and n sites per row. These sites are to be occupied by Ising spins (± 1). We assume that interactions exist only between nearest neighbors and that the value of these interactions are independent random variables : -J with probability c and J (>0) with probability 1-c. We also consider periodic boundary conditions (the torus) (refer to Fig. 1).

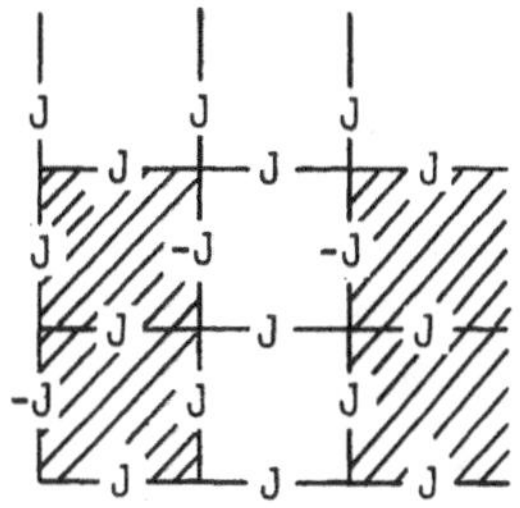

Fig. 1. A 3 × 3 spin glass. Dark plaquettes are "*curved*"

For a given set of interactions ($\{JxJ_{ij}\}$ with $J_{ij} = \pm 1$) and a spin configuration σ the energy will be of the form :

$$E(\sigma) = -J \sum_{(i,j)} J_{ij}\, \sigma_i\, \sigma_j$$

where (i,j) runs over nearest neighbor pairs of sites.

We recall briefly that an elementary square of the lattice is called a "plaquette". With each plaquette we associate the quantity $\Pi\, J_{ij}$ where the product is taken over the four bonds belonging to the plaquette. If this quantity is -1 the corresponding plaquette is said to be "curved" (refer to Fig. 1).

Without going into details [12], if we consider a sequence of $n \times n$ square lattices (n=1,2,...) with corresponding random partition functions :

$$Z(J^n,\beta) = \sum_{\sigma} \exp(\beta \sum_{(i,j)} J^n_{ij}\, \sigma_i \sigma_j)$$

the limit of the random free energy :

$$\frac{1}{n^2} \ln Z(J^n,\beta)$$

exists with probability 1.

If we denote this limit by $f(\beta)$ we have :

$$\lim_{n\to+\infty} \mathbb{E}\left(\frac{1}{n^2} \ln Z(J^n,\beta)\right) = f(\beta) \ .$$

1.2. Partition Functions: Polynomial Expressions and an Elementary Computing Algorithm

For any realization of random variable $Z(J^n,\beta)$ we can write :

$$Z(J^n,\beta) = z^{E_n}\, P_n(z) \quad , \quad z = e^{-2\beta}$$

where E_n is an integer and P_n is a polynomial.

The following experiments use polynomials corresponding to 6 x 6 square lattices. In order to compute the coefficients of such polynomials we use a method described in [5] ; we realize a formal summation for the states of the spins belonging to a particular subset of sites (decimation). Then, we enumerate the states of the other spins ; we change only one spin at each stage by using a Gray code.

1.3. Distributions of Zeros of Partition Functions: The Yang and Lee Theorem [1,6,13]

Under the hypothesis :

$$\lim_{n\to\infty} \frac{1}{n} \ln Z(J^n,\beta) = f(\beta)$$

it has been shown by YANG and LEE that any real singular point of f is an accumulation point of the set of the roots of polynomials P_n deduced from $Z(J^n,\beta)$. If we assume that the roots of these polynomials lie asymptotically on a regular curve, a critical (singular) point β_c is an intersection point of the previous curve with the real axis. In the case of the Ising Model (a spin glass with c=0) many arguments indicate that the roots lie asymptotically on two circles [2] (refer to Fig. 2).

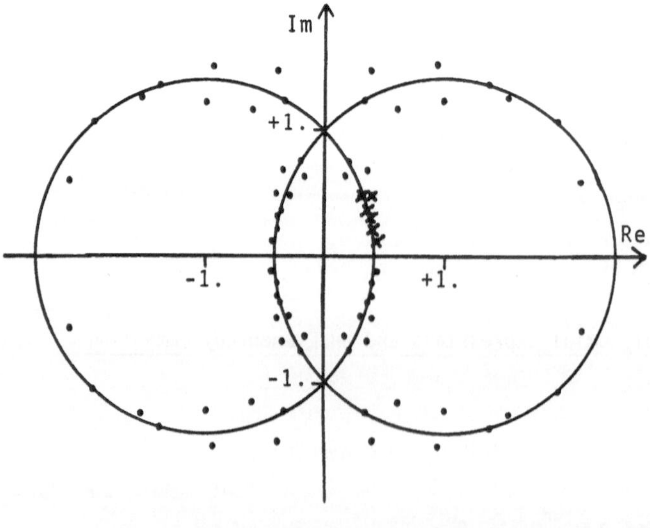

Fig.2. Distribution of roots on the $e^{-2\beta}$ plane for 6 × 6 Ising Model. (The points denoted by stars are used for the numerical computation of the critical value)

1.4. Superpositions of Samples

With the same value of the concentration c, we can average several independant realizations :

$$Z(J_1^n,\beta), Z(J_2^n,\beta), \ldots, Z(J_m^n,\beta)$$

of the same random variable. The multiplicative property :

$$\frac{1}{m} \sum_{i=1}^{m} \frac{1}{n} \ell n\, Z(J_i^n,\beta) = \frac{1}{mn} \ell n \prod_{i=1}^{m} Z(J_i^n,\beta)$$

justifies the superposition of the roots of all corresponding polynomials.

1.5. A Restricting Choice of Realisations

In a realisation of a $n \times n$ square lattice the observed value of concentration of negative interactions as well as the observed value of the concentration of curved plaquettes (the pertinent parameter in this problem [11]) are different from the theoretical values.

First, the results of many experiments have suggested to us that we select only the samples which realize the theoretical concentration of negative interactions. Then, in this first selection we tolerate a 10 % relative error on the concentration of curved plaquettes : in all observed cases this selection process reduces the fluctuation over an hypothetic limit curve (for example refer to Fig. 5 and Fig. 6).

We can give asymptotical justifications for the previous selection process. For small systems the improvement is only based on experimental results.

1.6. Numerical Results

To determine the critical value, $e^{-2\beta_c}$ with $\beta_c = J/kT_c$, we use a least square approximation by a parabolic curve passing through a set of points having small imaginary parts (for the Ising Model refer to Fig. 2).

The hypothesis of an asymptotic distribution of roots on a curve is plausible when the concentration c is smaller than 0.12; the critical values kT_c/J are plotted on the figure 3.

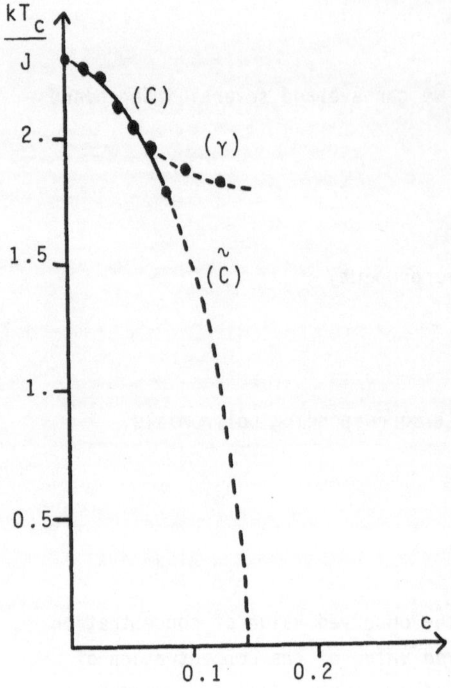

Fig. 3. Critical value $\frac{kT_c}{J}$ as a function of the concentration of negative interactions: c .

For the Ising Model the numerical value deduced from figure 2 is:

$$e^{-2\beta_c} = 0.423$$

which is a good approximation of the theoretical value $\sqrt{2}-1$.

When c is less than 0.08 the curve (C) is not surprising and we may extrapolate (C) by a curve $(\tilde{C})$ which predicts the absence of phase transition for c greater than 0.14.

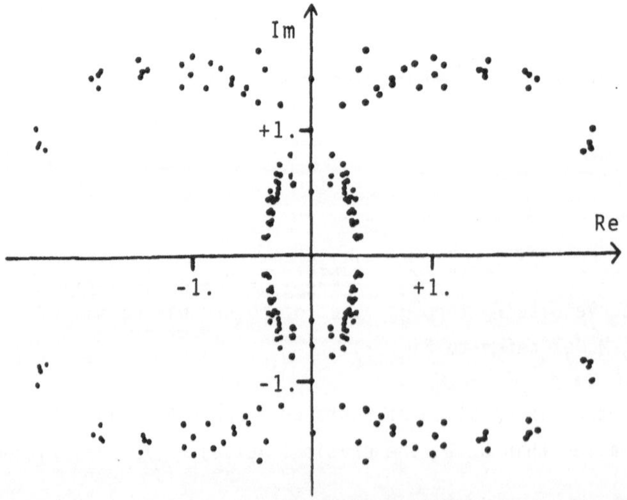

Fig. 4. Roots distribution (four samples) on the $e^{-2\beta}$ plane with c = 0.056.

Another experimental result is the appearance of a distinct curve (γ) above the previous curve. In the domain of high concentration the shape of distribution of roots is quite different (refer to Fig. 7).

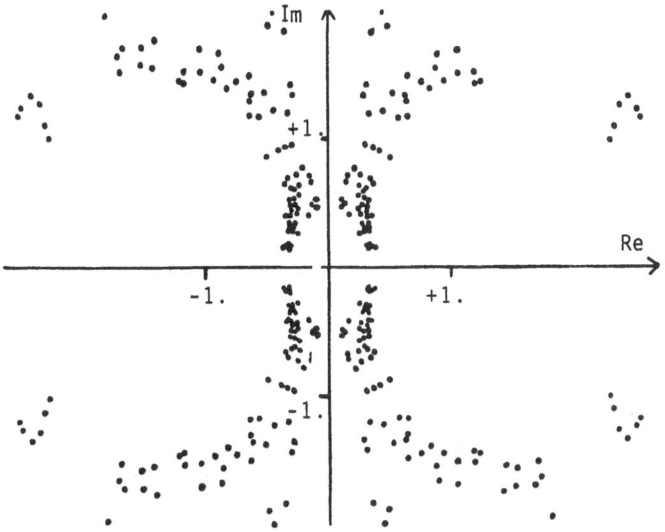

Fig. 5. Roots distribution (six samples) on the $e^{-2\beta}$ plane with c = 0.083 (without any restriction on the concentration of curved plaquettes).

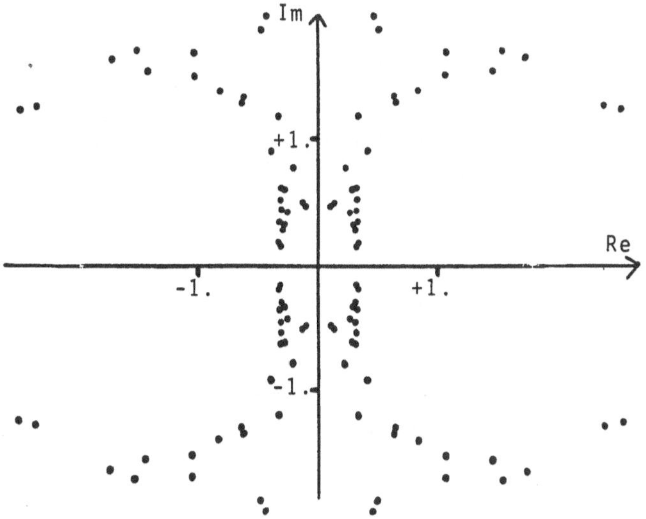

Fig. 6. Roots distribution (two samples) on the $e^{-2\beta}$ plane with C = 0.0.83 (concentration of curved plaquettes with 10% relative error).

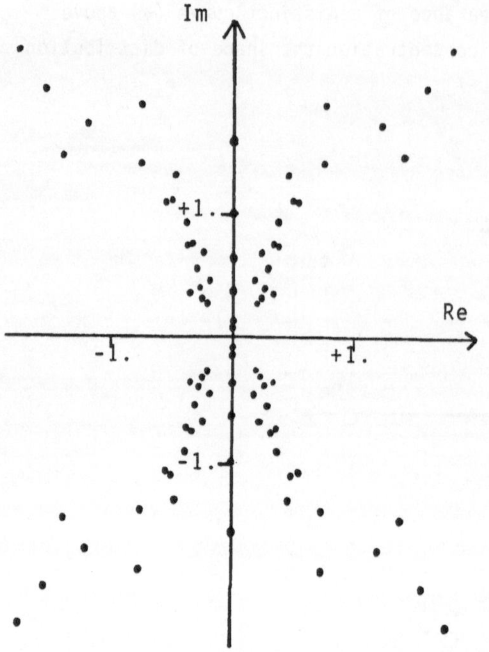

Fig. 7. Roots distribution (two samples) on the $e^{-2\beta}$ plane with c = 0.38

Part II Algebraic Method for the Computation of the Partition Functions of Spin Glasses

2.1 A Criterion of Complexity. Some Comparisons

We study the model described in 1.1 and we are interested in the computation, for a given value β (or e^{β} , $e^{-\beta}$) , of the quantity :

$$\sum_{\sigma} \exp(\beta \sum_{(i,j)} J_{ij}\, \sigma_i \sigma_j)$$

or in other form :

$$\sum_{\sigma} z^{\left(\sum_{(i,j)} J_{ij}\, \sigma_i \sigma_j\right)} \qquad z = e^{\beta} \ .$$

As a criterion of complexity we take the number of the elementary operations : additions and multiplications.

The direct enumeration of all the states of spins requires $0\,(n^2\, 2^{n^2})$ operations. With some improvements (formal summation : see 1.2) the previous cost decreases to $0(n^2\,2^{n^2/2})$. In this section we give an algebraic method which permits to decrease this cost to $O(n^2\,4^n)$. Furthermore this method is avalaible for many interesting examples.

2.2. Matrix Formulation of a Partition Function

Following Kramers and Wannier [4] we represent a configuration of spins by the set $\{\gamma_1,\gamma_2,\ldots,\gamma_n\}$ where γ_k is the configuration of the k-th row. The energy due to interactions within a row may be denoted $E_k(\gamma_k)$ and the energy due to interactions between two adjacent rows may be denoted $E_{k,k+1}(\gamma_k,\gamma_{k+1})$. We may define the following matrices :

A_k is a the diagonal 2^n-dimensional matrix, $A_k(i,i) = \exp\left(-\frac{\beta}{J}E_k(i)\right)$ $k=1,2,\ldots,n$

B_k is a 2^n-dimensional matrix , $B_k(i,j) = \exp\left(-\frac{\beta}{J}E_{k,k+1}(i,j)\right)$. $k=1,2,\ldots,n$

The partition function is [3,4,8] :

$$Z(\beta) = \text{trace}\,(A_1B_1A_2B_2\ldots A_nB_n) .$$

2.3. Expressions with Pauli Matrices

With Onsager and Kaufman [8, 3] we may note :

$$1 = \begin{pmatrix}1 & 0\\ 0 & 1\end{pmatrix} \qquad S = \begin{pmatrix}1 & 0\\ 0 & -1\end{pmatrix} \qquad C = \begin{pmatrix}0 & 1\\ 1 & 0\end{pmatrix}$$

$$S_r = 1 \otimes 1 \otimes \ldots \otimes S \otimes \ldots \otimes 1$$

$$C_r = 1 \otimes 1 \otimes \ldots \otimes C \otimes \ldots \otimes 1$$

with n factors in each direct-product $\otimes$, C and S appearing in the r-th position. Then we note :

$$\mathbb{1} = 1 \otimes 1 \otimes \ldots \otimes 1 \otimes \ldots \otimes 1 .$$

If $\overset{\circ}{J}{}^r_k$ denotes the sign of the interaction between the sites numbered r and $r+1$ of the k-th row, we have :

$$A_k = \prod_{r=1}^{n} ((\text{ch}\ \beta\ \overset{\circ}{J}{}_k^r)\ \mathbb{I} + (\text{sh}\ \beta\ \overset{\circ}{J}{}_k^r)\ S_r\ S_{r+1})$$

and if $\overset{\vee}{J}{}_k^r$ denotes the sign of the interaction between the r-th site on the k-th row and the r-th site in the (k+1)-th row, we have the equality :

$$B_k = \prod_{r=1}^{n} (\exp(\beta\ \overset{\vee}{J}{}_k^r)\ \mathbb{I} + \exp(-\beta\ \overset{\vee}{J}{}_k^r)\ C_r)\ .$$

Using these notations the evaluation of $Z(\beta)$ is equivalent to the computation of the trace of the product :

$$\prod_{\ell=1}^{2n^2} V_\ell$$

$$\text{with} : \quad V_\ell = \begin{cases} a_\ell\ \mathbb{I} + b_\ell\ S_r\ S_{r+1} \\ \quad\text{or} \\ a_\ell\ \mathbb{I} + b_\ell\ C_r \end{cases} \qquad (1)$$

where a_ℓ and b_ℓ are known complex numbers.

2.4. A Particular Basis of the Algebra of 2^n-dimensional Complex Matrix

We will used the basis constructed with the following matrices :

$$M_1 \otimes M_2 \otimes \ldots \otimes M_n \qquad M_i \in \{1, C, S, CS\}$$

and we note this basis by :

$$(\Gamma) = \{\Gamma_1 = \mathbb{I}, \Gamma_2, \ldots, \Gamma_{4^n}\}\ .$$

2.5. Property

We can determine two applications :

$$\varepsilon : \{1,2,\ldots,4^n\}^2 \to \{-1,1\}$$

$$f : \{1,2,\ldots,4^n\}^2 \to \{1,2,\ldots,4^n\}$$

which verify :

$$\Gamma_{k_1}\,\Gamma_{k_2} = \varepsilon(k_1,k_2)\,\Gamma_{f(k_1,k_2)} \qquad \forall\, k_1,k_2 \in \{1,2,\dots,4^n\}\ .$$

In order words the product of two elements of the basis (Γ) is an element of the some basis, except for the sign. We assume that the computations of $\varepsilon(k_1,k_2)$ and $f(k_1,k_2)$ require no elementary operations.

Proof

We notice that the set $\{1,C,S,CS\}$ is stable by multiplication, except for the sign. Moreover we have :

$$(M_1 \otimes M_2 \otimes \dots \otimes M_n)(M_1' \otimes M_2' \otimes \dots \otimes M_n') = (M_1M_1') \otimes (M_2M_2') \otimes \dots \otimes (M_nM_n')$$

this proves 2.5.

2.6. Property

Using the notations of 2.3, if :

$$\prod_{\ell=1}^{L} V_\ell = \sum_{i=1}^{4^n} \gamma_L^i\, \Gamma_i$$

where the γ_L^i are known , and if we note :

$$\prod_{\ell=1}^{L+1} V_\ell = \sum_{i=1}^{4^n} \gamma_{L+1}^i\, \Gamma_i$$

we can compute the set of 4^n values : γ_{L+1}^i with less than 3×4^n elementary operations.

Proof

Matrix V_{L+1} may be expressed (see 2.3) :

$$V_{L+1} = a\,\Gamma_1 + b\,\Gamma_m$$

and then :

$$\prod_{\ell=1}^{L+1} V_\ell = \Big(\sum_{i=1}^{4^n} \gamma_L^i\,\Gamma_i\Big)(a\,\Gamma_1 + b\Gamma_m) = \sum_{i=1}^{4^n} (a\,\gamma_L^i)\Gamma_i + \sum_{i=1}^{4^n} (b\,\gamma_L^i)\,\Gamma_i\,\Gamma_m$$

and we deduce the expression :

$$\sum_{i=1}^{4^n} (a\,\gamma_L^i)\Gamma_i + \sum_{i=1}^{4^n} (b\,\gamma_L^i)\,\varepsilon(i,m)\,\Gamma_{f(i,m)} \qquad .$$

The computations of the set of values γ_{L+1}^i requires $2.\,4^n$ multiplications and 4^n additions .This proves 2.6 .

2.7. Final Property

$Z(\beta) = \text{trace}\,(A_1B_1 \ldots A_nB_n) = 2^n\,\gamma^1_{2n^2}$ and the evaluation of $\gamma^1_{2n^2}$ require $O(n^2\,4^n)$ elementary operations.

Proof :

We have : trace $(\Gamma_1) = 2^n$ and trace $(\Gamma_i) = 0$, $i \neq 1$ and the product (1) involves $2n^2$ factors.

2.8. Storage Requirement

The previous method requires an amount of storage wich growes as $O(4^n)$.

2.9. Some Generalizations

The same method can be used to evaluate the partition function of a spin glass with a magnetic field :

$$Z(\beta,\gamma) = \sum_\sigma \exp\,(\beta \sum_{(i,j)} J_{ij}\,\sigma_i\sigma_j + \gamma \sum_i \sigma_i)$$

by insertion of factors : $\text{ch}\,\gamma\,\mathbb{I} + \text{sh}\,\gamma\,S_r$.

It is possible to apply this method to other graphs of sites with most general interactions and also to three dimensional models ; but in this last case the complexity factor is most important.

About the non-periodic two dimensional model the complexity factor decreases to $O(n^2\,2^n)$: some experiments with different algorithms were realized recently [7]. In the same way experimental algorithms are used in the "Laboratoire de Mathématiques Appliquées de Grenoble".

Bibliography

1 DOMB, GREEN : Phase transitions and critical phenomena Vol. 1

2 M.E. FISHER : Lectures in theoretical physics 7c (University of Colorado Press, Boulder, 1965).

3 B. KAUFMAN : Crystal statistic II. Partition function evaluated by Spinor analysis.
Phys. Rev., Vol. 76, Number 8 october 15, 1949.

4 H.A. KRAMERS and G.H. WANNIER :
Phys. Rev., 60, 252-263 (1941).

5 B. LACOLLE : Le modèle Verre de Spins.
Rapport de Recherche n° 182. Laboratoire de Mathématiques Appliquées de Grenoble.

6 T.D. LEE, C.N. YANG :
Statistical theory of equation of state and phase transition II. Lattice gas and Ising Model Phys. Rev., Vol. 87, n° 3, august 1952.

7 I. MORGENSTEIN, K. BINDER
Magnetic correlations in two-dimensional spin glasses. Preprint.

8 L. ONSAGER : Crystal statistics I. A two-dimensional model with an order-disorder transition. Phys. Rev. 65, 117 (1944).

9 S. ONO, Y. KARAKI, M. SUZUKI, C. KAWABATA
Statistical thermodynamics of finite Ising Model I. Journ. of Phys. Soc. of Japan, Vol. 25, n° 1 (1968).

10 M. SUZUKI, C. KAWABATA, S. ONO, Y. KARAKI, M. IKEDA :
Statistical thermodynamics of finite Ising Model II. Journal of Phys. Soc. of Japan, Vol. 29, n° 4 (1970).

11 J. VANNIMENUS, G. TOULOUSE :
Theory of the frustration effect. Ising spin on square lattice.
J. Phys. C 10 Solid State Physics, L537.

12 P.A. VUILLERMOT Thermodynamics of quenched random spin systems and application to the problem of phase transition in magnetic (spin) glasses.
J. Phys. A. : Math. Gen, Vol. 10, n° 8, (1977).

13 C.N. YANG, T.D. LEE :
Statistical theory of equations of state and phase transition I. Theory of condensation.
Phys. Rev. Vol. 87, n° 3, august 1952.

Percolation and Gelation by Additive Polymerization

P. Manneville and L. de Sèze
Dph-G/PSRM, CEN-SACLAY, BP 2
F-91190 Gif sur Yvette, France

It has long been well known that branching polymers in solution can undergo a transition from a liquid state, the sol phase, to a gel phase characterized by the existence of an infinite macromolecule [1]. In this paper we shall consider the case of additive polymerization where the molecules are growing by free radical reactions:

$$R^{\bullet} + R' - CH = CH - R'' \rightarrow R - \underset{\displaystyle R'}{\underset{|}{CH}} - \underset{\displaystyle R''}{\underset{|}{C^{\bullet}H}} \ .$$

An important parameter is the functionality of the monomers which is the number of other molecules that can be linked to them. In our case the functionality is just twice the number of double bonds which can be opened by free radical reactions.

Once the process is initiated by introduction of free radicals in the system (initiators), the molecules grow and simultaneously the fraction p of functions used for linking increases. If the solution contains a fraction of tetrafunctional monomers, the molecules cross-link and we can observe, at a given threshold p_c, a transition characterized by a divergence of the viscosity of the solution and the onset of an elastic behavior. This threshold corresponds to the apparition of an infinite macromolecule, the gel, among the finite molecules, the sol.

The classical theory of gelation was introduced 40 years ago by FLORY and STOCKMAYER (FS) [2,3]. They identified p with a uniform probability for any function of a monomer to be used for linking with <u>any other one</u>, neglecting the occurence of cycles and steric hindrance effects. GORDON et al. [4] have improved FS work with the use of cascade theory, predicting with reasonable accuracy the gel points of various systems.

In 1976 STAUFFER [5] and de GENNES [6] underlined the importance of the cycles and steric hindrance effects, especially in the vicinity of the gel point and proposed an analogy between gelation and percolation. The polyfunctional units are placed at the nodes of a lattice and p is identified with the probability that any function of a monomer is used for linking with a <u>neighboring unit</u>. If the functionality of all units is equal to the number of first neighbors for the given lattice one recovers a standard bond percolation model (Fig.1a).

In this model, thresholds are lattice-dependent and thus theory cannot correctly account for the observed values but the behavior in the vicinity of the gel point is characterized by critical exponents (Table I) which depend only on space dimensionality d and are very different for d=3 from those predicted by FS theory (Table II). In fact FS theory can be interpreted in terms of percolation on a Bethe lattice (Fig.1b) and corresponds to a mean field theory of the gelation. Its exponents are expected to be exact only for $d \geq 6$.

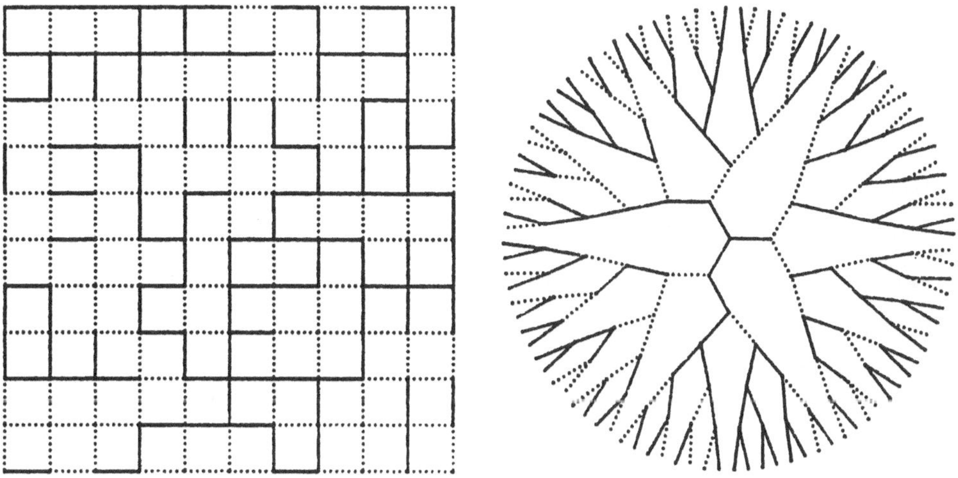

(a) (b)

Fig.1. Examples of bond percolation samples on a square (a) and a Bethe lattice (b)

Heavy lines indicate closed bonds (linked units) which are randomly chosen with a probability p.
Clusters (molecules) are formed by nodes (monomers) joined by closed bonds.
Fig.1b shows clearly that the Bethe lattice (used in FS theory) cannot be fully developed in finite space dimensionality, due to steric hindrance effects.

Extending the analogy to "special" percolation models, de GENNES related the viscosity below the gel point to the conductivity of a random network of super- and normal- conductors [7] and the elasticity above the gel point to the conductivity of a random resistor network [6].

The analogy with percolation has clearly shown the importance of cycles and steric hindrance effects in gelation but several important questions are still open. STAUFFER [5,8,9] and de GENNES [10] have already discussed the problems of the introduction of a lattice, the absence of motion of the units and the extent of the critical region around the threshold where significant differences with FS predic-

Table I Analogies between gelation and percolation and definition of the critical exponents [5].
In the vicinity of a critical point p_c, a physical quantity X has a singular part of the form $(p-p_c)^x$ where x is the critical exponent associated to X. The table gives the correspondence between the physical quantities of interest in gelation and percolation and the usual symbols for the associated critical exponents.

Gelation	Percolation	critical exponent
conversion factor p	bond probability p	
gel point p_c	percolation threshold p_c	
number of macromolecules N	number of clusters N	$2-\alpha$
gel fraction G	percolation probability P	β
degree of polymerization DPw	mean cluster size	$-\gamma$
correlation length	correlation length	$-\nu$
elasticity	conductivity (random resistor network)	t

Table II Numerical values of critical exponents for classical FS theory of gelation and standard 3D percolation

	β	γ	ν	t	σ	τ
FS	1	1	1/2	3	1/2	5/2
3D percolation	0.4	1.7	0.84	1.7	0.48	2.15

The exponents σ and τ have been defined by STAUFFER [5,12] and their relation to the other exponents is given in the text.

tions should be observed.

As another question about the relevance of standard percolation results for gelation, one can wonder whether a uniform random drawing of bonds can generate statistical distribution of molecules which will be representative of those obtained in a real growth process, different distributions leading to different critical exponents. To clarify this point, we found it interesting to perform simulations of additive polymerization on a lattice.

The Model and the Framework for Analysis of Data

We have chosen to study the gelation process of a solution of bifunctional (2fu) and tetrafunctional (4fu) units in absence of any solvent. Units (monomers) are placed at the nodes of a simple cubic lattice. The different states of the units are symbolized on Fig.2.

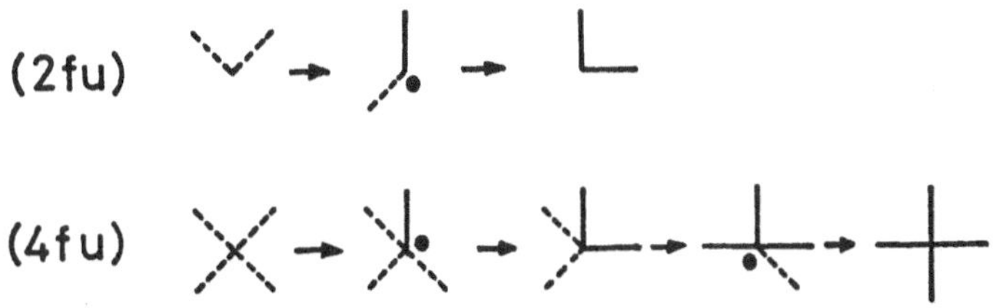

Fig. 2 : Different states of 2fu and 4fu.

- dotted lines indicate functions available for linking
- heavy lines indicate functions used for linking with a neighboring unit
- dots indicate that the unit is an "active center" i-e has a free-radical able to open a double bond of a neighboring unit and to link with it.

Before starting the simulation, the 4fu are distributed randomly in the sample with a concentration c_{4fu}, the other nodes being occupied by 2fu. The process is initiated by activation of a given concentration c_I of randomly chosen units.

At each step :

i) one of the active centers is selected at random ;

ii) one of its eighteen first or second neighbors having available functions is selected at random ;

iii) the two units are linked ;

iv) the active center is transferred to the neighbor or annihilated with the active center of the neighbor.

The "normal" death for an active center is the encounter with another active center, but another case occurs when an active unit has no neighbor with available function. This accident is clearly related to the absence of motion of the units in our model ; however the possibility of linking with second neighbors keeps this event quite rare.

We defined the gel phase by the existence of a cluster with units in all the horizontal planes ("infinite" cluster). The sample is then said to be connected.

For each sample we determined the gel point. The cluster size distribution and the number of cycles were collected at regular intervals. All results were studied as a function of the concentration p of "nodes" i-e the fraction of the 4fu with at least 3 functions used for linking. We checked that no singularity occurred for the relation between p and the number of steps (which represents time in the process), even in the vicinity of the gel point. To analyse the results of our simulations we used the standard methods developed for percolation simulations [11].

1. Analysis of the percentage of connected samples (pcs) for different sample sizes (Fig.3a)

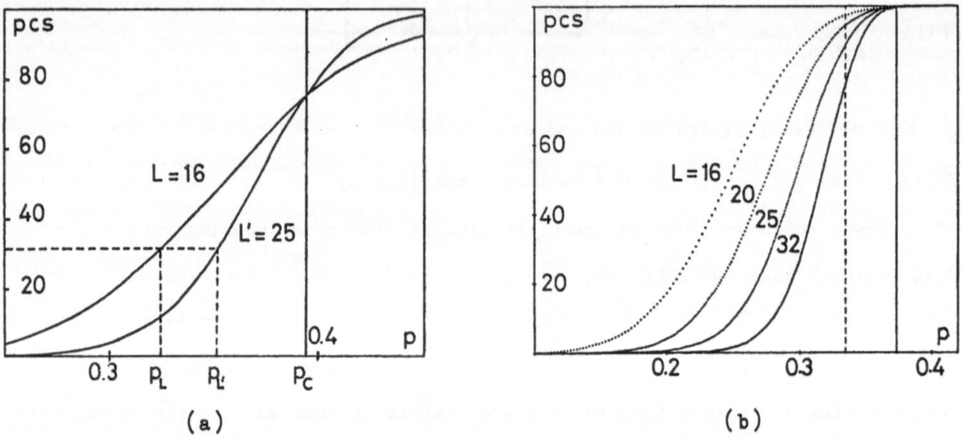

Fig. 3 : Percentage of connected samples (pcs) as a function of the fraction p of cross-links for concentrations of tetrafunctional units c_{4fu} = 0.2 and c_{4fu} = 0.1 and different sample sizes.

In Fig.3a the intersection of the curves gives an estimation of the threshold p_c. p_L and $p_{L'}$ at the same pcs are related through the scaling relation :
$pcs(p,L) = f[(p_c-p)L^{1/\nu}]$.
In Fig.3b the heavy line indicates the threshold $p_c = 0.37$, estimated directly from the analysis of these curves. The dashed line indicates the threshold $p_c = 0.33$. estimated from cluster size distribution analysis, using the standard 3D percolation values for the critical exponents.

If we assume that the correlation length ξ is the only characteristic length near the gel point, a same pcs will be observed for samples of different linear sizes L and L' if the correlation lengths are in the ratio L/L' :

$$\xi_L(p_L)/\xi_{L'}(p_{L'}) = L/L' .$$

This relation defines the threshold as the fixed point of the transformation :

$$\xi_L(p_c)/\xi_{L'}(p_c) = L/L' .$$

Since ξ behaves near the threshold as $(p_c-p)^{-\nu}$ we can postulate the following scaling form for the pcs in the vicinity of the gel point [11] :

$$\text{pcs}(p,L) = f[(p_c-p)\, L^{1/\nu}] .$$

2. Cluster size distribution

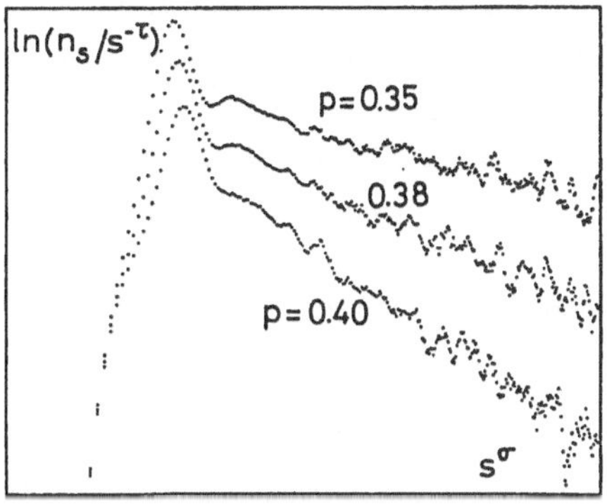

Fig. 4 : Cluster size distribution as a function of the number of sites for different values of $p(c_{4fu} = 0.1)$.
The unusual aspect of the curves for small cluster sizes is due to the very special structure of the clusters in this model.
These curves are related through the scaling relation $n_s = q_o s^{-\tau} f[q_1(p-p_c)s^{\sigma}]$.

In 1975 STAUFFER proposed a scaling form for the number of clusters of s sites near the threshold [12,5] :

$$n_s = q_o s^{-\tau} f[q_1(p-p_c)s^{\sigma}] \text{ with } f(x) = \exp(-x) \text{ above } p_c .$$

The exponents τ and σ are related to the usual critical exponents :

$$2 - \alpha = \frac{\tau-1}{\sigma} \quad , \quad \beta = \frac{\tau-2}{\sigma} \quad , \quad \gamma = \frac{3-\tau}{\sigma} .$$

Fig.4 shows typical plots of $\ln(n_s/s^{-\tau})$ versus s^{σ}. For large s the distribution becomes unreliable due to the fact that the dimension of the clusters is of the order of the sample size. We used periodic boundary conditions in order to minimize these effects.
We also examined the "percolation probability" i.e. the probability for a unit to be in the infinite cluster, in order to determine the exponent β but we found the accuracy of the results and the size of the samples not sufficient to obtain valuable estimations.

Results of the Simulations

We used 4 different sample sizes :

$16 \times 16 \times 16$: 15500 samples

$20 \times 20 \times 20$: 10000 samples

$25 \times 25 \times 25$: 11000 samples

$32 \times 32 \times 32$: 6500 samples .

Remaining parameters are the 4fu concentration c_{4fu} and the initiator concentration c_I. In order to approach experimental conditions one has to decrease c_{4fu} as much as possible. But doing this one meets two conflicting conditions about c_I. Indeed c_I controls the average length of chains in absence of cross-linking. When c_I is small, chains are long and a spurious connectedness can be observed in small samples. When c_I is large, chains are shorter and at a given c_{4fu}, the average number of 4fu/chain may be below the value necessary for gelation to occur so that not all samples will reach the gel point.

We found a compromise for $c_I = 0.1$ and were thus able to decrease c_{4fu} down to 0.1.

The structures of the clusters we observed are very different from those obtained in standard percolation. Two facts are very symptomatic of this difference :

- the pathological aspect of the cluster size distribution for small s(Fig.4) which is due to the existence of "closed" small clusters which cannot be merged into large clusters due to the lack of available functions.
- the coexistence of several distinct "infinite" clusters beyond the gel point. This feature is, of course, observed less and less as the sample size increases. However this fact implies a ratio perimeter/size very different from standard cases.

"High" 4fu concentration

For $c_{4fu} = 0.2$ the pcs curves are quite similar to those obtained for standard percolation (Fig.3a). The exponent ν is found to be $\nu = 0.84 \pm 0.04$ agreement with standard percolation value.

"Low" 4fu concentration

For $c_{4fu} = 0.1$ almost all samples are connected in the vicinity of the gel point and the accuracy of the data is very poor (Fig.3b). We found $\nu = 0.9 \pm 0.1$ and $p_c = 0.37$. Thus ν seems larger than in standard 3D percolation but the error bar does not allow to decide whether this difference is significative.

At first sight the situation is still worse for the cluster size distribution where we have to determine three parameters τ, σ and p_c. In order to obtain a clear-cut answer we determined p_c, using the values $\tau = 2.15$ and $\sigma = 0.48$ of 3D percolation (Table I). We found $p_c = 0.33$ which is definitely not compatible with pcs results (Fig.3b).

Using the pcs threshold $p_c = 0.37$ and $\sigma = 0.48$ we obtain $\tau = 2.35$. If we introduce these values in the hyperscaling relation $d\nu = 2-\alpha = (\tau-1)/\sigma$ we recover the value $\nu = 0.9$ obtained from pcs curves analysis.

From this consistency argument we can conclude that the deviation of the value we observed for ν at low concentration is not accidental.

Conclusion

The exponent ν we determined cannot be directly related to experimental measurements but we can use a relation proposed by de GENNES [6] between ν and the exponent t characteristic of the behavior of the elasticity above the gel point :

$$t = 1 + \nu(d-2)$$

Such a relation, which is in reasonable agreement with the results of simulations for standard 3D percolation, indicates that to our larger value of ν should correspond a larger value of t. This prediction should be compared with the results of M. ADAM [13] who has measured the elasticity of styrene and divinylbenzene above the gel point and obtained a value of $t = 2.1$ much larger than the standard 3D percolation value $t = 1.75$.

Our model suffers from two drastic simplifications : introduction of a lattice and absence of motion of the molecules. Nevertheless it clearly demonstrates that taking into account the growth aspect of the process along with steric hindrance effects leads to molecule structures rather different from those obtained in the standard percolation case and even farther from those of the classical tree-like model.

Moreover the coexistence of large entangled clusters can justify serious doubts about the analogy between the mechanical properties of gels and the electrical properties of random networks. As a matter of fact those clusters, though not linked chemically, are certainly knotted together (trapped entanglements). This should affect the mechanical properties in a way ignored by the electrical analogy.

If standard percolation does not seem well-adapted to the description of additive polymerization gelation, the question remains open for other types of gelation such as polycondensation. Simulations on corresponding models should be useful to clarify the situation.

Acknowledgments

The authors are indebted to P.G. de GENNES and M. ADAM for drawing their attention to this subject and for many discussions during the course of this work and to D. STAUFFER for communication of preprints.

References

[1] P.J. Flory, Principles of Polymer Chemistry, Cornell University Press, Ithaca (1953).

[2] P.J. Flory, J. Am. Chem. Soc. 63, 3083 (1941).

[3] W.H. Stockmayer, J. Chem. Phys. 11, 45 (1943), 12, 125 (1944).

[4] M. Gordon, Proc. Roy. Soc. London, A268, 240 (1962).

[5] D. Stauffer, J. Chem. Soc., Faraday Trans. II, 72, 1354 (1976).

[6] P.G. de Gennes, J. Phys. Lettres (Paris), L37, 1 (1976).

[7] P.G. de Gennes, C.R. Hebd. Séan. Acad. Sc. , B286, 131 (1978).

[8] D. Stauffer, 1980 Prague Meeting on Macromolecules, Polymer Networks, Karlovy Vary, Sept. 1980. To be published in Pure and Applied Chemistry.

[9] D. Stauffer, STATPHYS 14, Edmonton, Alberta, Canada, Aug. 1980.

[10] P.G. de Gennes, Scaling Concepts in Polymer Physics, Cornell University Press, Ithaca (1979).

[11] S. Kirkpatrick, Lecture Notes, Les Houches Summer School on Ill Condensed Matter (1978) North Holland.

[12] D. Stauffer, Phys. Rev. Letters 35, 394 (1975), Physics Reports, 54, 1 (1979).

[13] M. Adam, 1980 Prague Meeting on Macromolecules (see Ref.8).

Ground State Structure of the Random Frustration Model in Two Dimensions

R. Maynard and R. Rammal

Centre de Recherches sur les Très Basses Températures, CNRS, 166X
F-38042 Grenoble Cédex, France

1. Introduction

The problem of spin glasses is of growing interest both in solid state physics and in statistical physics. This problem emerged from the observed anomalous properties of dilute magnetic alloys, such as 1% of Mn or Fe embedded in Cu or Au. The most striking property of these systems is the cusp in the magnetic susceptibility at a well defined temperature. From this observation the question of a new phase (spin glass) at low temperature arose; this would be distinct from the canonical ferro magnetic phase (standard symmetry breaking).

In a simple model, we can use Ising spins S_i on a lattice for which values can be only ± 1. It is assumed that interaction J_{ij} between spins can have only two symmetrical values ± J. The energy associated to this system is given by the expression

$$H = - \sum_{i} J_{ij} S_i S_j \quad . \tag{1}$$

The distribution of J_{ij} is of the form

$$P(J_{ij}) = x\, \delta(J_{ij} + J) + (1 - x)\, \delta(J_{ij} - J) \tag{2}$$

$0 \leq x \leq 1$. It is believed that despite its strong simplifications, this model retains the relevant features of real spin glasses. In the absence of transparent physical picture of the new phase, it is of importance to understand the low temperature behaviour of this system. Particular attention is paid here to the study of ground states (minimize H, with respect to spins, given J_{ij}'s) of this frustration model (i.e. at T = OK), but when x varies from x = 0 (ferromagnetic ground state) to x =.5 (the spin glass state)(the phase diagramme being symmetric around x = .5, on a square lattice).

The threshold x^*, above which the ferromagnetism disappears is studied by numerical simulation, using Edmond's algorithm. The determination of this value, together with the characterization of ground states in terms of clusters of solidary spins (cluster of spins which can be reversed simultaneously without any cost in energy), of living bonds (rigidity of ground states) and of defect lines or fractures (which run from one side to the other on a finite sample permitting the spin of a large portion of the sample to be reversed) are the main objects of the following sections.

2. Failure of the Monte-Carlo Algorithm

The study of the ground states in the frustration model has been tentatively pursued in the past, by the well-known relaxation method [1]. A Monte-Carlo step consists in choosing a spin at random, calculating the energy change ΔE associated with flipping this spin and flipping the spin with this probability: $\min(1, \exp(-\Delta E/T))$. However it has been realized that this relaxation method can give neither the thermodynamic equilibrium states, nor the ground state for the frustration model.

In order to explain this failure let us consider the problem of the relaxation at low temperature T of the well-known 2 d Ising model (or the Mattis model of spin glasses). From an initial random configuration of spins, the system evolves towards the ground state by shrinking of the domains of wrong oriented spins. The average time of coalescence is proportional to n, the size of the domain, while the energy of this domain varies as n, the perimeter, [2]. In this case, the excited states decrease monotonically to the ground state without energy barriers.

The situation in frustration model (and generally in spin glasses) is radically different. When we compare first excited states to a ground state, it appears clearly that a cluster of spins of relatively large size must be flipped at some time to fall down in the ground state. All these clusters of solidary spins can be classified in two distinct categories.

Entropic clusters which can be reversed spin by spin at constant energy (except the last, where the energy decreases to the fundamental energy) following some very precise paths in phase space).

Pinned clusters which contain at least one spin the flip of which costs energy - the potential barrier -. When this spin is reversed, all the others of the cluster follow without any cost of energy.

These clusters can overlap each other : the structure of these states is not a simple partition of the set of spins. Moreover, there is a strong correlation between these clusters, since the nature of one cluster can be modified by reversing the state of the neighbour overlapping cluster. Then, these clusters are not independent and the model is not reducible to some sort of superparamagnetism.

The origin of the failure of standard relaxation methods resides in these clusters of spins [3]. The study of their dynamic and particularly the life-time of excited states, shows the following behaviour.

Entropic clusters: The life time corresponding to a size cluster n is $\tau \sim n^2$. This result reflects the strong degeneracy of first excited states, and contrasts with the corresponding behaviour $\tau \sim n$ in the regular Ising model.

Pinned clusters: The life time in this case is given by the expression $\tau \sim n^2 \exp(n_b \Delta E/T)$ at low temperature. Here n_b denotes the number of energy barriers of height ΔE.

These results give a deeper insight to the failure of the Monte-Carlo algorithm. We believe that the above discussion remains at $d \gtrsim 2$, and reflects in general the strong metastability of spin glasses: high potential barriers and high degeneracy.

3. The Edmond's Algorithm

A new procedure was used to overcome the difficulties of the Monte-Carlo method [4]. This method maps the frustration problem into the problem of finding a matching of maximum cardinality and minimum weight in the graph of frustrated plaquettes. In this way, the use of Edmond's algorithm [5] permits the generation of the ground states in a relatively short computation time. In the following, we reformulate the ground state problem in a more direct way than [4], permitting using carefully the pertinent variables of the problem: the frustrated cycles.

3.1. Formulation of the Ground States Problem

Let us consider a graph $G = (S,E)$, where each vertex i is assigned a spin $S_i = \pm 1$. With each pair of spins S_i and S_j, which are the ends of the bonds $e = (i,j) \in E$, we associate an interaction energy of the form $- J_{ij} S_i S_j$, where J_{ij} denotes a given real number. The energy associated to this system is given by (1). It is not hard to see that, for every cycle C of G, containing an odd number of negative assigned edges, there exists no configuration of spins in C such as every bond has its minimum energy. In this case, the cycle C is said to be frustrated. An edge having its minimum energy is to be called satisfied. Every frustrated cycle C contains at least one frustrated bond, in any ground state.

Examples

(i) On a simple square lattice, it is easy to see that the set of all elementary square cycles, with four edges - plaquettes - contains all information on the frustration contents of the lattice.

(ii) If we impose periodic boundary conditions to the square lattice, it is needed to take into account at least two additional plaquettes (superplaquettes).

3.2. Primal-Dual Problem

We limit ourselves here to planar lattice. Let denote by C a frustrated cycle (odd set). To each edge $e = (ij)$ in the graph G, we associate a real number X_{ij} so that

$$X_{ij} = \begin{cases} 1 \text{ if } (ij) \text{ is violated} \\ 0 \text{ otherwise} \end{cases} .$$

The problem of the ground state (i.e. minimize H with respect to $S_i = \pm 1$, for all i, given a configuration of J_{ij}'s) can be formulated as follows

$$(P) \begin{cases} \min \sum_{(ij)\in E} x_{ij} |J_{ij}| \\ \sum_{e\in C} x_e \geq 1 \text{ for any frustrated cycle } C \\ x_e \geq 0 \text{ for any edge } e \in C \end{cases} .$$

This formulation is a direct consequence of the work of Edmond and Johnson on the Chinese postman problem [7].

The dual problem associated with (P) is (D)

$$\left.\begin{array}{l} \max \sum_{c} Y_c \\ \left\{ c \quad \sum_{(ij)\in C} Y_c \leq J_{ij} \text{ frustrated cycle} \right\} \\ Y_c \geq 0 \end{array}\right\} (D)$$

(D) represents the distribution of energy between frustrated cycles (see figure 1).

Any distribution Y satisfying the constraints of (D) gives a lower bound for the energy of frustrated cycles:

$$\sum Y_c \leq \sum J_{ij} X_{ij} .$$

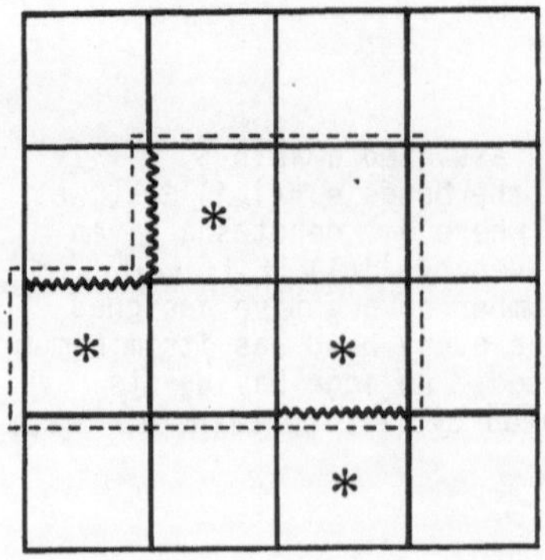

Fig. 1. Example of frustrated square lattice.
——— positive bond; ∿∿∿ negative bond;
* frustrated plaquettes;----frustrated cycle.

The best lower bound: max $\Sigma\, Y_C$ is exactly the minimum energy of violated edges: min $\Sigma\, |J_{ij}|\, X_{ij}$.

Using the weak theorem of complementary slackness [8], we can derive the following criterion for the optimality of primal and dual solutions.

Optimality criterion. A pair (X,Y) of solutions for (P) and (D) are optimals if and only if

$$X_{ij} > 0 \text{ implies } \sum_{\{C|(ij)\in C \text{ frustrated cycle}\}} Y_C = J_{ij}$$

and

$$Y_C > 0 \text{ implies } \sum_{(ij)\in C} X_{ij} = 1 \quad .$$

Consequences (a) if $\Sigma\, Y_C < |J_{ij}|$, then (ij) cannot be violated in any ground state

(b) if $Y_C > 0$, the frustrated cycle C contains one and only one frustrated edge in any ground state.

3. 3. Rigidity - Living Bonds

Instead of using the degeneracy properties of ground states, it is more relevant to use the concept of solidary spins [9], to study the long range order in the ground state. The packet of solidary spins is defined as a group of spins that keep the same relative orientation in all ground states. These packets provide a measurement of the rigidity.

The rigidity can be characterized by the set of rigid bonds (in contrast to living bonds). A rigid bond is by definition a violated bond or non-violated, for all the ground states.

To find the set of rigid bonds, we use the following procedure. Starting with a ground state (solution X_0 of (P)), a violated bond (resp. non violated) is a living bond if and only if, for any $\varepsilon > 0$ added (resp. substracted) to its weight $|J_{ij}|$ we destroy the optimality of X_0.

This post-optimal analysis can be down with a polynomial algorithm, using the primal algorithm of the chinese postman problem. In fact, the algorithm of rigidity has O (N^2) as complexity [10], where N denotes the total number of spins.

The practical implementation of this procedure, as well as the detailed algorithm, cannot be described here. But it is interesting to make some remarks concerning this algorithm.

Remark 1. In contrast to the Monte-Carlo technique, the algorithm described here is not based on a relaxation procedure. During the course of the computation, we have a solution for spin configuration at each step, in contrast with matching algorithm used previously [4]. As opposed to relaxation methods, our procedure permits the flipping of spins set having any complicated form (non local algorithm). In this way we can overcome the metastability difficulties.

Remark 2. The complexity of this algorithm is polynomial, as the matching one. But in contrast to the latter, the post-optimal analysis gives the set of rigid bonds. To our knowledge, this is the first algorithm able to give in a non ambiguous way the set of rigid bonds in the frustration model.

Remark 3. Finally, it is important to recall the result concerning the change of algorithmic complexity, when a 3d lattice is used instead of a planar one. In a 3d lattice, the problem becomes NP- complete [10], but for any planar lattice, we have a polynomial one.

4. Numerical Results

A numerical study has been performed over a set of 243 different square lattice of size varying from 10 x 10 to 20 x 20, and concentration of negative bonds x, varying in the range $0.10 < x < .50$, as well as at .50. The boundary conditions used are not standard ones. Matching edges have been introduced in order to have the standard torus configuration in the spin lattice, but in the lattice of plaquettes (dual lattice), the set of superplaquettes has been ignored [6]. The right picture of these boundary conditions is the repeated cell scheme or a tesselation of the plane. We term these as pseudo-periodic boundary conditions. In practise, only two boundary conditions can be found: pseudo-periodic where the spins of opposite sides are parallel, and pseudo-antiperiodic where the spins of opposite sides are antiparallel.

This additional degree of freedom permitted us to exhibit apparition of magnetic walls structure for moderate values of x.

4.1. Rigidity - Fracture Lines

Using the rigidity algorithm described above, two situations are found:
- Large cluster of rigid bonds, extending between the four sides (percolating rigidity or rigid samples).
- Small finite clusters of rigid bonds only (fracturated samples).

For each concentration x of negative bonds, the average fraction of rigid bonds F has been calculated. The decrease of F (x) describes the pulverisation of the infinite rigid cluster (present at x = 0) into finite clusters. Therefore, the long range order existing at low concentration disappears above a critical concentration $x_c \sim .15$, where a new short range order phase takes place. The analysis of this behaviour, using the concept of fracture lines, has been done. These lines are defined [4] as alternating cycles running through the sample from one side to the other. The probability of occurence of these lines versus x is shown in figure 2. The same value $x_F = x_c \sim 0.15$, obtained reveals the proliferation of the fracture lines, and the absence of long range rigidity, at high concentration ($x > x_F$).

It is interesting to remark that the percolation of rigid bonds occurs here at a critical concentration of rigid bonds, around 0.7, instead of .5, the threshold of uncorrelated bond percolation problem.

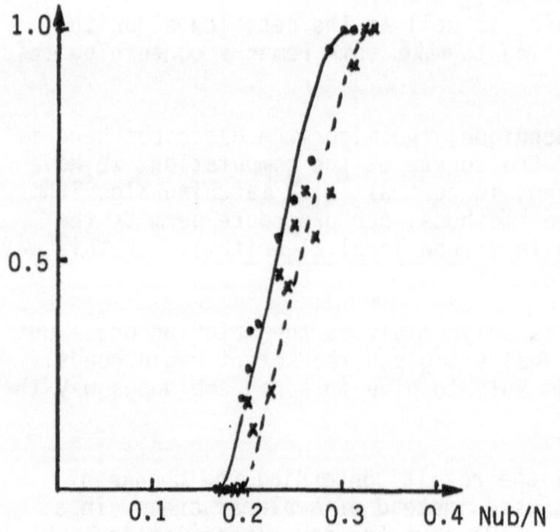

Fig.2. Frequency of occurence of fracture line (x) (243 samples) and magnetic walls(o) (173 samples) versus the number of unsatisfied bonds Nub. This number is proportional to the ground state energy E_o (per spin): $E_o = -2\ (1\text{-Nub/N})$

4.2. Random Antiphase State

Below $x_F \sim .15$, where the samples are rigid, two types of ground states are observed.
- ferromagnetic ground state, below $x_m \sim .10$
- a random antiphase state, for $x_m < x < x_F$.

The random antiphase state is structured in magnetic domains, separated by magnetic walls. The probability of occurence of these magnetic walls has been calculated and plotted in fig. 2. The threshold which can be deduced from this figure corresponds to x_m. The observation of this new random antiphase state is unexpected: its interpretation in terms of spin glass phase is in progress.

5. Conclusion

The standard Monte-Carlo method fails to generate low energy or ground states in frustration model. The origin of this failure comes from the existence of large clusters of spins which must be reversed at the same time to generate ground states. The mean life time varies as n^2 or $n^2 \exp(\alpha n)$ instead of n in case of homogeneous phase.

A simple polynomial algorithm known in graph theory as the Edmons's algorithm has been successfully applied to solve this problem. It processes by reversing packets of solidary spins of any size and shape. A determination of the threshold of occurence of macroscopic fracture lines is performed : $x_F \sim .15$. For $x > x_F$, only small finite clusters of rigid bonds exist, there is no long range order and no spin glass phase. The percolation problem of rigid bonds gives a critical concentration of rigid bonds around .7.

A new phase called random antiphase state is observed in the range $x_m < x < x_F$, where $x_m \sim .10$ and $x_F \sim .15$. In this phase, the ground states are characterized by an infinite cluster of rigid bonds, structured in antiphase domains, separated by magnetic walls.

Acknowledgements

The authors acknowledge warmly F. BARAHONA and J.P. UHRY from the department of Applied Mathematics of the University of Grenoble, for their initiation to graph theory and the active cooperation on the application of the Edmond's algorithm to the frustration problem as well as the numerical experimentation.

References

[1] K. Binder(ed.), Monte-Carlo Methods in Statistical Physics, Topics in Current Physics, Vol. 7 (Springer, Berlin, Heidelberg, New York 1979)
S. Kirkpatrick, Phys. Rev. B16, 4630 (1977)
[2] C.P. Yang, Amer. Math. Soc. 15, 351 (1963)
[3] R. Rammal, R. Suchail, R. Maynard, Sol. State Comm. 32, 487 (1979)
[4] I. Bieche, R. Maynard, R. Rammal, J.P. Uhry, J. Phys. A. Math. Gen. 13, 2553 (1980)
[5] J. Edmonds, Can. J. Math. 17, 449 (1965)
[6] F. Barahona, R. Maynard, R. Rammal, J.P. Uhry (to be published)
[7] J. Edmonds, E. Johnson, Math. Progr. 5, 88 (1973)
[8] E.L. Lawler, Combinational Optimization : Networks and Matroids (New York : Holt, Rinehart and Winston, 1976)
[9] J. Vannimenus, J.M. Maillard, J.L. De Seze, J. Phys. C : Solid St.Phys. 12, 4523 (1979)
G. Toulouse, Commun. Phys. 2, 115 (1977)
[10] F. Barahona, thesis (unpublished), University of Grenoble (1980).

Line Defects and the Glass Transition

N. Rivier and D.M. Duffy
Blackett Laboratory, Imperial College
London SW 7, U.K.

Abstract

The structure of glasses contains line defects, characterized by oddness rather than intensity. The density of these defects in equilibrium, above a given temperature T_o where they are free to move, is proportional to the fluidity of the supercooled liquid and follows a simple function of the temperature known empirically as the VOGEL-FULCHER law. The maximal entropy per mole associated with the defects is of order of R ln 2, in good agreement with experimental data on melting. Below T_o, the defects are frozen in and the glass is an elastic solid.

1. Structure of glasses

Structurally, glasses can be divided into two broad classes : 1) Continuous random networks [1], in which bonds between the constituting, non spherical atoms or molecules can be identified naturally. Window glasses (SiO_2), amorphous Ge or Si, are standard examples, but spin glasses and the network of pores in a porous medium like a rock also belong to this class of structures. 2) Amorphous packings [2], where the atoms can be randomly but closely packed, so long as the boundary conditions are not too regular. Amorphous and liquid metals, even, to some extent, metglasses (made of transition metals and metalloids in the proportion 3 : 1 to 4 : 1) are typical examples, even if the latter exhibit covalent short-range order around the metalloid atoms [3].

One type of structure is the dual of the other. Consider, for example, a random packing of spheres. It is possible to construct around every sphere a VORONOI polyhedron, which is the analogue in glasses of the WIGNER-SEITZ cell in crystals. The continuous network made of the edges, vertices of the polyhedra and the polyhedra themselves forms an unambiguous network of pores [4], which is, by construction, the dual of the sphere packing structure. When the spheres all pack in tetrahedral cells (this is not possible if all the spheres are identical) [5], all pores are four-functional units, and the porous network has fixed coordination number 4.

Random structures have no constructive symmetries, and only that regularity which allows every vertex to be as suitable a reference point as any other. This is often labelled local invariance and is mathematically analogous to gauge invariance. At first sight, glasses look as dull, complicated and structureless as a phone book, and contemplation of either seems equally likely to induce sleep. (This constitutes an empirical demonstration of global translational symmetry or stationnarity).

Yet, it has been shown recently [6], [7] that all random structures have specific, topologically stable line defects, which are a particular kind of disclination (rotation dislocation) and are their own antidefects. Thus there are not one (the matter density) but two (the density of disclinations as well) quantities following conservation laws in glasses. The same situation occurs in liquids where density (point) and vorticity (line) are two conserved fields. Consequences of the presence and of the motion of these line defects will be investigated here. Firstly, disclinations tend to pair up and lower their elastic energy at low temperatures, while the pairs dissociate at high temperatures where the gain in entropy more than offsets the energy loss in the free energy. We shall see that for topological, line defects, even in 3 dimensions, there is a precise balance between energy and entropy and the system has a phase transition similar qualitatively to the KOSTERLITZ-THOULESS transition in two-dimensional systems [8]. The viscosity and other relaxation times associated with the motion of dissociated line defects follows accordingly the VOGEL-FULCHER law [9][10]. Secondly, since a liquid is a disordered structure and a crystal is ordered, an essential part of the entropy difference on melting will be associated with the defects and is topological.

2. Line defects in glasses (disclination, frustration)

The identification of the line defects is based on the

Theorem [6]. Continuous lines, closing as loops or terminating on the surface of the glass, can be threaded through all odd rings of bonds, avoiding even rings.

Proof : Let every edge i carry the weight $J = -1$. For every ring of bonds (face) α, associate the product $\phi = \Pi_{i\in\alpha}(J)$. Then $\phi = \pm 1$ for an even/odd ring. Any closed surface S on the network will have

$$\Pi_{\alpha\in S}(\phi) = \Pi_{i\in S}(J)^2 = 1 \quad (1)$$

since every edge separates exactly two rings on S. Thus every S contains an even number of odd rings, providing an exit for every entering disclination line.

The defects have been called disclinations in glasses since, if one starts from a crystalline, diamond structure with 6-fold rings, odd rings are constructed by cutting, rotating the lips, and regluing. The disclinations are characterized by oddness rather than by intensity. They are their own antidefects and are labelled by the two-elements group Z_2. In spin glasses, these lines thread through frustrated plaquettes. In random packings, the defects are closed chains of odd edges (every edge being the apex of an odd number of faces). This follows immediately by duality from the unambiguous identification of defect lines in the (covalent) porous network. (In principle one could also identify another dual family of line "defects", odd edges in the covalent network, odd faces in the sphere packing. Apart from the ambiguity in their identification, these "defects" are not dynamical variables in the usual cases, where the coordination number of the network. and thus all edges, have a definite parity (in SiO_2, $z = 4$. In other networks edges disappear and vertices coalesce, but the parity of z remains invariant), or the number of faces per cell of the random packing, and thus all faces, have constant parity (if all cells are tetrahedra [5], or even canonical BERNAL polyhedra [2], all faces are triangular). These "defects" are an invariant part of the chemical structure, they are either nowhere or everywhere, and their number cannot be changed by physical means).

It is possible to construct random structures without any such defects. These structures are called trivial or flat, because the disorder can be eliminated by a gauge transformation. (MATTIS model of spin glasses, CONNELL-TEMKIN continuous random network, but a defect-free random packing has yet to be constructed).

The existence of stable Z_2 line defects in disordered condensed matter can be proved in general using homotopy theory. One starts with a local field theory, in which the field $\Delta : \mathbb{R}^3 \rightarrow \mathcal{M}$ is a mapping of continuous, ordinary space onto a manifold of internal states $M = G/H$. Here G is the group of all possible transformations in the Lagrangian, whereas $H \subset G$ is the isotropy subgroup of the field. M contains all possible ground states, as well as all excitations of the systems, whether non-topological (spin waves or phonons) or topological (defects), i.e. whether or not the image in M of a closed contour in $\mathbb{R}^3$ is homotopic to zero. Hence, a defect is topologically stable if it can be surrounded by a non-trivial contour (in direct generalization of GAUSS', AMPERE's or BURGERS' theorems), that is if the corresponding homotopy group π_n (M) is non-trivial [11] .

In ordered condensed matter, Δ is the local order parameter. In spin glasses or glasses, it cannot be simply a spin or tetrahedron because we are faced with an essential difficulty. The interaction between spins or covalent atoms is defined on a network rather than on the continuum $\mathbb{R}^3$. It is not defined below a given, finite distance (the bond length or sphere radius), and cannot be in general be extrapolated to infinitesimal distances, unless the network is such that there is never any competition between interactions anywhere. (In that case, a gauge transformation gives the same sign to interactions which can then be extrapolated to infinitesimal distances. The model is flat). In response to the competition between interactions, the spins (or tetrahedra) compromise and take up non-colinear (or non-staggered) arrangements (Fig. 1).

To absorb the interaction into $\mathbb{R}^3$ we must take as local field a spin configuration spanning a small but finite region, which is necessarily non-collinear. The field is therefore a hedgehog, a shaggy (hirsute) configuration of spins or tetrahedra, described locally by the full rotation matrix, and $\mathcal{M} = SO(3)$, rather than by the tetrahedral group or $\mathcal{S}^2 = SO(3)/SO(2)$ as in the case of single spins [7] .

Consequently, in glasses, G is the Euclidean group $G = SO(3) \wedge T(3)$ (including translations) and the isotropy group $H = 1 \wedge 1$ is trivial since there is no order of the local fields. Thus $\mathcal{M} = G$ and $\pi_1(\mathcal{M}) = Z_2$. The only stable line (n = 1) defects are disclinations, and the local field is rotated by 2π - like a spinor - when it is taken around a closed contour surrounding the disclination.

Our argument yields a limiting distance a below which the theory is invalid, a core size for the defect, and a corresponding upper limit for the density of defects. Screening of the energy of defects requires a lower limit for this density (see also Section 3.2), so that we are in a semi-dilute regime [12] . This screening is done by the defects themselves, in the following fashion :

Topological line defects can also give rise to more conventional, non-topological distortions. For example, a hairpin disclination loop made of dipoles of 5 and 7 fold rings constitutes a segment of an edge dislocation (Fig. 2).

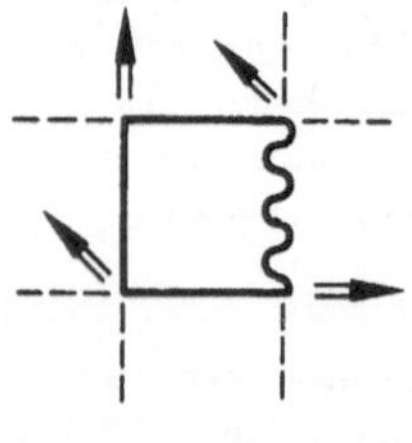

Fig. 1 Competition of interactions in spin glasses (frustrated plaquette) with resulting non-collinear spins. The wiggly (straight) line represents an antiferro (ferro) magnetic interaction.

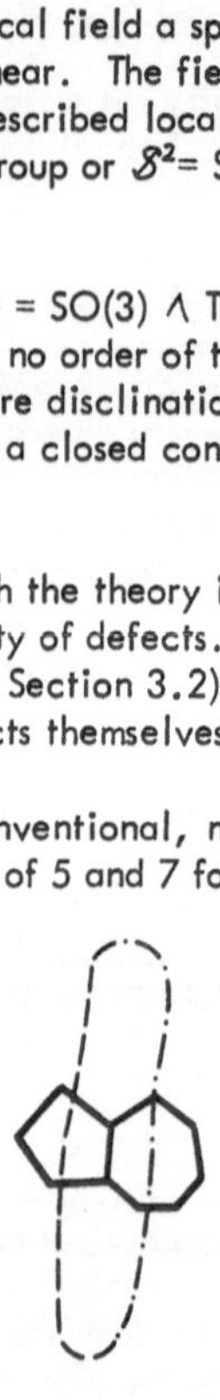

Fig. 2 The hairpin disclination loop minimizes locally the elastic energy in the glass. It constitutes a segment of dislocation.

The edge dislocation is not topologically stable in glasses, and the intensity or sign associated with 5 or 7 rings is not a topological charge, but locally we have two local rotations about different, parallel axes, that is a translation perpendicular to the axes, since translations form a normal subgroup of the full Euclidean group. These segments of dislocation screen the stresses due to disclinations. Radiolarias use this technique to build a honeycomb skeleton on a sphere, requiring 12 5-fold rings, whose stress is screened by many dislocations or 5-7 dipoles. A similar screening mechanism is also observed in BENARD cells (J. PANTALONI, private communication).

3. Relaxation in supercooled liquid

It is universally found that the rate of all transport processes of supercooled liquids decreases rapidly, but smoothly, near the glass transition temperature, T_g. It has the following functional form which has become known as the VOGEL or FULCHER Law[9]:-

$$R = (\text{const})\ e^{-A/(T-T_0)} \tag{2}$$

[9], [10]. R represents the fluidity, inverse relaxation time, diffusion rate, nuclear relaxation rate and the rate of all other processes of the supercooled liquid. The constants A and T_o are the same for all processes of a given material, and T_o generally lies between 10°c and 180°c below T_g. The universality of the law (2) (which is known under various names in different areas of research, polymers, oils, etc.) suggests that all transport processes are dominated by a common relaxation mechanism, with timescale divergent at T_o, hence only below T_o is the glass a truly elastic solid.

The defect diffusion model of relaxation in liquids is based on the assumption that a site (or group of sites) can only relax to a state of lower energy when a defect has succeeded in diffusing to it [13] [14]. The defect, as emphasised by ANDERSON [10], is not simply a hole or free volume without permanent identity, but it must satisfy a conservation law. A disclination line would contribute to the relaxation process via the distortion it imparts to the surrounding structure. This increases the potential energy of the neighbouring sites and enables them to cross the energy barriers and take a configuration of lower energy.

Consider a medium with a length of defect for unit volume given by ρa^{-2} then the average distance between the molecular sites of the medium and the nearest defect will be proportional to $\rho^{-1/2} a$.

Assuming that the defects diffuse by a random walk process, then the average time necessary for a defect to diffuse to a site is proportional to the square of the distance between them, i.e.

$$\bar{\tau} = R^{-1} \sim a\rho^{-1} \quad . \tag{3}$$

Thus the assumption that a site relaxes only when a defect diffuses to its neighbourhood, leads directly to an average relaxation time inversely proportional to the density of defects in the medium. The density of defects in equilibrium at a given temperature can be found by minimizing the free energy,

$$F(\rho) = E(\rho) - TS(\rho) \quad ; \quad \left.\frac{\partial F}{\partial \rho}\right)_{eq.} = 0 \quad . \tag{4}$$

3.1 Entropy of Disclination Loops

Since it is often stated that the entropy per unit length of a line defect should be a constant, rather than a logarithmic function of the density of defects, [10], we shall give first an outline of the physical situation which yields a logarithmic entropy.

There are two contributions to the entropy of a set of L loops with a total of K elements :

i) A configuration entropy $S_{conf} \sim K \ln (z-1)$, where z is the coordination number of the network, and

ii) An entropy of mixing which is, for small loops, $S_{mix} \sim L \ln L$, because small loops can be counted as point elements. For large loops, as in the case of polymers in solution, $S_{conf} \gg S_{mix}$, and the entropy per unit length is indeed a constant, whereas, for small loops, $S_{mix} \gg S_{conf}$ and the entropy is logarithmic. There are many more ways of splitting a fixed number K of elements into small loops than into long ones, hence the small loop dominates the entropy, $K \sim L$ and $S \sim K \ln K$.

Explicit calculation of the entropy goes as follows. The number of configurations of K disclinated or odd faces forming l loops in a continuous random network of F faces, when $1 \ll K \ll F$, is given by

$$\Omega = \sum_{\ell=1}^{L} \binom{F}{\ell} \sum_{n_1 \ldots n_e} \frac{C(n_1)}{n_1} \cdots \frac{C(n_e)}{n_e} \, \delta\Big(\sum_{i=1} n_i - K\Big) \tag{5}$$

Here, the combinatorial factor counts the possible starting points of the l loops. Each loop with n odd faces has C (n) configurations, the number of closed self-avoiding walks of n steps. It has also n equivalent starting points, hence the factors 1/n. Since $K \ll F$, exclusion of the faces already occupied by previous loops yields a negligible correction.

A precise calculation of C (n) even though it is unknown is unnecessary, as mixing of small loops dominates the entropy. For definition, we set

$$C(n) = \varkappa \, n^{-\beta} \, \zeta^{n} \tag{6}$$

by analogy with a well-tested scaling law for self-avoiding walks [15] , with the connective constant $\zeta \lesssim z-1$ (the equality being reached in networks where the self-avoiding condition is negligible), $\beta \gtrsim 1 + 7/4$ in three dimensions, and $\varkappa \sim \frac{1}{2}$.. Defining a generating function for a single loop,

$$f(x) = \sum_{p=n_0}^{\infty} \frac{x^p}{p^\beta} = \frac{x^{n_0}}{n_0{}^\beta} \, g(x) \tag{7}$$

with n_o the number of elements in the smallest possible loop, we obtain

$$\Omega = \frac{\zeta^K}{K!}\left(\frac{\partial}{\partial x}\right)^K \left\{[1+f(x)]^F - 1\right\}_{x=0} \simeq \frac{\zeta^K}{K!}\left(\frac{\partial}{\partial x}\right)^K [e^{F\varkappa f(x)} - 1]_{x=0} \tag{8}$$

in the limit $F \gg 1$. The sum over l in (5) has been carried over to ∞, since the lowest order term in f^{ℓ}, $x^{\ell n_o}$, has vanishing K'th derivative at $x = 0$ whenever $l > L \sim K/n_o$. Letting $K = n_o L + r$, with $0 \leqslant r \leqslant n_o - 1$ and L, the maximum possible number of loops, both integers, and expanding the exponential, we have

$$\begin{aligned}\Omega &= \zeta^K \sum_m \frac{(\varkappa F/n_0^{\beta})^m}{m!(K-mn_0)}\left(\frac{\partial}{\partial x}\right)^{K-mn_0} g^m(x)\,|_{x=0} \\ &= \zeta^K \frac{1}{L!}\left(\frac{\varkappa F}{n_0^{\beta}}\right)^L \frac{1}{r!}\left(\frac{\partial}{\partial x}\right)^r g^L(x)\,|_{x=0}[1+0(F^{-1})] \quad .\end{aligned} \tag{9}$$

But, since $K > L \gg n_o > 1$,

$$\frac{1}{r!}\left(\frac{\partial}{\partial x}\right)^r g^L(x) = \frac{1}{r!} L^r\left(1+n_0^{-1}\right)^{r\beta}[1+0(L^{-1})] \quad . \tag{10}$$

Its logarithm is of order ln L and its contribution to the entropy negligible. In terms of the dimensionless disclination density $\rho = K/F$, the entropy is

$$S = k_B \ln\Omega = F[D\rho - C\rho\ln\rho] + 0(\ln K) \tag{11}$$

where $C = L/K$ is a positive coefficient of order 1, and $D = \ln\zeta - C\ln\left[C n_o^{\beta}/(\varkappa e)\right]$, is also at most of order 1, and is only logarithmic in the geometric parameters $\varkappa, \beta, \zeta^o$ of the network. (Eqn. (6)).

3.2 Energy of Disclination Loops

We are in the semi-dilute regime, where there is only one length scale in the problem [12]. The density of disclination is not high enough, so that their cores do not overlap, no self-avoiding potential energy need be included, the structure of the glass and its distortions are given by a single local field describing the hedgehogs of atoms introduced in section 2. But the density is still sufficiently high to enable the disclinations to screen each other's elastic energy, most efficiently through the hairpins of Fig. 2. The strain field of a disclination line is screened and falls off as the inverse of the distance [16], instead of being independent of the distance from the line, as in crystals. The situation is the same as in the hexatic and liquid phases of two-dimensional melting [16] or in liquid crystals [17].

The system is therefore essentially a collection of topological defects, and the main screening mechanism originates from the defects themselves. There is only one length scale in the problem, the average distance between disclinations ξ. Like the entropy the energy is therefore topological (with a scale determined by a bulk elastic modulus of the glass), and the microscopic details of the structure or of the interatomic forces are not important.

The topological character of the line defects implies that

$$\oint_{\Gamma} \underline{A}(\underline{x})d\underline{x} = \text{finite} \quad ; \quad \oint_{\Gamma} \underline{A}(\underline{x})d\underline{x} = 0 \tag{12}$$

for any contour Γ surrounding one single defect. Here A is some projection of the local field. The Z_2 algebra of topological intensities in (12) suggests choosing for the manifold of the internal states $\underline{A}$ (x), a real projective plane of dimension n > 1. The planar, or XY model n = 1 would be suitable if algebra modulo 2 was assumed independently. Eq. (12) is valid within the scale $a < |\underline{x}| < \xi$, where a is the size of the core of the disclination or the bond length.

The situation described by (12) is similar, apart from the algebra modulo 2, to the electromagnetism of current loops. That is, the energy of disclination loops i with linear elements $\underline{dl}$ is

$$E = \frac{1}{2} X \sum_{ij} \int \frac{d\ell_i \cdot d\ell_j}{r_{ij}} \tag{13}$$

similar to the self- or mutual induction of a system of current loops. Here r_{ij} is the distance between the linear elements dl_i and dl_j, and $X = \mu a^2/4\pi$, where μ is the shear modulus, is a measure of the stiffness of the glass. For all loops geometries [17] [18],

$$E = Xa \sum_{i=1}^{K} \ln\left(\frac{r_i}{a}\right) + E_{core} \tag{14}$$

where r_i is the distance between the ith segment and the nearest antiparallel defect. There are K segments of dislocations of length a in the glass. A core energy has been added.

For random configurations of defect loops, we should average (14) over a probability distribution P(r) of distances r. This distribution is a function of the dimensionless variable $y = r/\xi$ since the only scale of length of the problem is $\xi = a\rho^{-1/2}$, where $\rho = K/F$ (The density of line defects $\mathcal{L}/\mathcal{V} \sim Ka/(Ca^3)$ has dimension (length)$^{-2}$, and $C = (2/s)F$ where s is the average number of faces per cell). We obtain

$$E = F[B\rho - A\rho\ln\rho] \tag{15}$$

where $A = \frac{1}{2}Xa$ and $B = E_{core} + Xa\langle \ln y\rangle$ with $\langle \ln y\rangle = \int dy\, P(y) \ln y / \int dy\, P(y)$. Eq. (15) has the same form as the entropy (11), with only the coefficient of ρ depending on the particular structure of the glass.

The algebra modulo 2 does not affect (13) or (15). A direct, microscopic calculation of the energy of one frustration loop in a XY spin glass on a lattice [19] yields also (15).

3.3 Equilibrium Density of Free Defects

Substituting (11) and (15) into the expression (4) for the free energy, we obtain

$$F(\rho) = F[(CT-A)\rho\ln\rho + (B-DT)\rho] \tag{16}$$

and the equilibrium density of free defects, or, by (3), the relaxation rates, are given by

$$R \propto \rho_{eq} = \alpha \exp[-\gamma/(T-T_0)] \qquad (17)$$

with $\alpha = \exp(D/C - 1)$, $\gamma = (B - DA/C)/C$ and $T_o = A/C$.
γ is always positive, because the core energy is larger than the bulk elastic energies, but only the latter may be affected by external constraints. $k_B T_o \simeq \mu a^3$ is simply the shear elastic energy. Eq. (17) is the VOGEL-FULCHER law observed experimentally.

Hence the density of free defects falls continuously but rapidly to zero at some temperature T_o. The defects loops do not, in general, annihilate by shrinking, but they form tightly bound pairs of disclinations, or bend into hairpin configurations (Fig. 2). Both configurations have the strain field of segments of dislocations. Notice the close similarity of the transition at T_o (in three dimensions), with the two-dimensional KOSTERLITZ-THOULESS transition [8]. The essential steps in our derivation have been i) line defects, ii) semi-dilute regime (self-screening by the defects), and iii) the existence, without stiff penalty in energy, of small defect loops. A similar situation to glasses may occur in very inhomogeneous superconductors, where the weak links caused by the inhomogeneities of the structure form a random network and small vortex loops may surround even an individual weak link (R. VOSS, private communication).

There are several alternative derivations of the VOGEL-FULCHER law. The most famous is the free-volume theory, but it is not overridingly convincing (cf [10] for references and criticisms). The present derivation, which follows a suggestion of ANDERSON [10] is sufficiently general and model-independent to encompass such a general law as (17). The simplicity of this, single-length scale derivation suggests that deviations from the law (17) - which occur in the temperature range of most supercooled liquid - are due to departure from the semi-dilute regime. This, at any rate, may be tested experimentally, as is the dependence of the parameters of (17) on external constraints like pressure.

4. Entropy Change on Melting

It has been suggested recently [20] that a similar, empirical law to TROUTON's law of vaporization existed for melting, in that the volume independent change in entropy on melting equals R ln 2 per mole, where R is the molar gas constant. This law has been confirmed experimentally for a large class of simple metals [21]. The latent entropy is therefore the sum of a contribution due to the change of volume on melting, and a universal quantity which should be directly related to the different topologies of a crystal and its corresponding liquid. We shall show that the topological entropy can be accounted for by the presence of line defects (the disclinations of Section 2), moving freely in the liquid and absent in the crystal. (This may suggest an indirect method of measuring the zero point entropy of glasses as the discrepancy between the entropy of the liquid, extrapolated to the same volume as the glass at $T = 0$, and R ln 2).

In crystals, line defects are absent, or at least negligible in comparison to the liquid phase, where disclinations are free to move, and can be created spontaneously, so that the change of topological entropy on melting is the configurational and mixing entropy of an arbitrary number of free Z_2 defects.

Consider first a random, covalent network, made of N vertices (atoms), E edges (bonds), F faces (rings) and C cells. We shall also need its average coordination number z. The number of configurations of an arbitrary number of free defects in the network is simply two configurations per face (free to be odd or even), under the constraint that the lines threading odd faces must be continuous. The constraint is imposed by keeping one face in each cell as a parity control, thus providing an exit, if necessary, for a line entering that cell. The number of independent faces is therefore F - C, and the topological entropy S_t of the liquid is given by $S_t = k_B(F-C)\ln 2$. The EULER-SCHLAEFLI formula for a 3 dimensional network, $N-E+F-C=0$, yields

$$S_t = k_B(E-N)\ln 2 = k_B N\left(\frac{1}{2}z-1\right)\ln 2 \tag{18}$$

because $E = \frac{1}{2}zN$. For continuous, tetrahedrally bonded random networks, $z = 4$ (at each vertex, not on average) and $S_t = k_B N \ln 2$, that is $R \ln 2$ per mole.

Similarly, at first sight, for random close packed structures, each edge can be odd or even, and the continuity of the line defect is guaranteed by keeping one edge per vertex as a parity control, so that an even number of odd edges always meet at every vertex. Thus, for a given random structure (N^*, E^*, F^*, C^*; z^*),

$$S_t^* = k_B(E^*-N^*)\ln 2 = k_B N^*\left(\frac{1}{2}z^*-1\right)\ln 2 \quad . \tag{19}$$

For random packings of identical, hard spheres, $z^* = 12-14$, [2] and S_t^* is about 6 times higher than the experimental value. Clearly, use of identical elements impose considerable steric constraints on the random packing, since the distance between vertices and the angle between incident edges has strict lower bounds. These constraints, and the correct topological entropy, can be evaluated by the following argument : the parity of an edge (ij) can only be changed if an edge (ik) on one of the incident faces ceases to exist, i.e. if the contact between the spheres i and k is interrupted. (The interruption also creates a few additional small loops). This interruption is possible without modification or relaxation of the structure, only if the spheres i and k remain fully constrained, i.e. have at least twice as many contacts per sphere as they have degrees of freedom (3). It follows that at least six edges per vertex must be uninterrupted. A packing at rest must be iso - or hyperstatic [22] . This steric constraint fixing 6 edges per vertex replaces the topological constraint of continuity of the line defect involving only 1 edge per vertex, and we obtain, instead of (19), for identical spheres

$$S_t^{id*} = k_B(E^*-6N^*)\ln 2 = k_B N^*\left(\frac{1}{2}z^*-6\right)\ln 2 \simeq kN^*\ln 2 \tag{20}$$

since $z^* = 14$ (defined unambiguously as the number of faces per cell in the porous network [4]). The steric restrictions are considerably weakened if one uses spheres of various sizes, and the topological entropy should increase from $R\ln 2$ to $\lesssim 6R\ln 2$ (cf (19)) in multicomponent liquids.

The porous (covalent) network is, as we have seen, the dual of the close packed network. The steric constraint above, based on the interruption of contact as the main agent of creating disclinations, corresponds in the porous network to the coalescence of pores as the main modifier of ring parity. This is as observed in BENARD cells. (J. PANTALONI, private communication)

Incidentally, duality between the two networks, with $N = C^*$, $E = F^*$. $F = E^*$ and $C = N^*$, implies equality of their topological entropies,

$$S_t = S_t^* \quad . \tag{21}$$

Both entropies can be written in terms of topological invariants. The covalent network can be regarded as a graph in a 3 dimensional Euclidean space, characterized by the cyclomatic number (the first Betti number of the graph) $R_1 = E - N + 1 = N(\frac{1}{2}z - 1) + O(1)$, so that in the thermodynamic limit,

$$S_t = kR_1 \ln 2 \quad . \tag{22}$$

Moreover, $R_1 = R_1^*$ is a particular case of ALEXANDER's duality theorem [23] in d = 3, and thus so is eq. (21), which has been obtained above by elementary arguments.

5. Conclusions and Acknowledgements

After identifying line defects in disordered condensed matter we have shown that they can account for some of the common properties of glasses and liquids,

i) the VOGEL-FULCHER law for the relaxation rates or the fluidity as a function of temperature in viscous and supercooled liquids, and
ii) the topological entropy difference between liquid and solid states.

NR is grateful to the staff of the Département de Physique des Systèmes Désordonnés of the Université de Provence, Marseille, where part of this work was done, for many enlightening discussions and comments, and particularly to J. Pantaloni for showing him films and photographs of his experiments on the Rayleigh-Bénard instability prior to publication. He also acknowledges valuable discussions with P. Lacour-Gayet and J.A. Dodds, as do both authors with M. Warner and N.R. da Silva. D.M.D. acknowledges the support of an S.R.C. Studentship.

References

1. D.E. Polk, J. non-cryst. Solids 5, 365 (1971)
2. J.D. Bernal, Proc. Roy. Soc. A 280, 299 (1964)
3. P.H. Gaskell, J. Phys. C 12, 4337 (1979)
4. J.L. Finney, Proc. Roy. Soc. A 319, 479 (1970)
5. J.A. Dodds, Nature 256, 187 (1975); J. Coll. I. Sci., to be published
6. N. Rivier, Phil. Mag. 40, 859 (1979)
7. G. Toulouse, Phys. Reports 49, 267 (1979)
8. J.M. Kosterlitz and D.J. Thouless, J. Phys. C 6, 1181 (1973)
9. R.H. Doremus, Glass Science (Wiley 1973), ch. 6
10. P.W. Anderson, in Ill-Condensed Matter, R. Balian, R. Maynard and G. Toulouse, eds. (North Holland 1979), ch.3, section 2.1
11. G. Toulouse and M. Kléman, J. Physique Lettres 37, L 149 (1976)
12. M. Daoud et al, Macromolecules 8, 804 (1975)
13. S. Glarum, J. Chem. Phys. 33, 639 (1960)
14. M.C. Phillips, A.J. Barlow and J. Lamb, Proc. Roy. Soc. A 239, 193 (1972)
15. C. Domb, Adv. Chem. Phys. 15, 229 (1969)
16. B.I. Halperin and D.R. Nelson, Phys. Rev. B 19, 2457 (1979)
17. P.G. de Gennes, Liquid Crystals, (Oxford 1974), section 4.2
18. L.D. Landau and E.M. Lifshitz, Electrodynamics of Continuous Media (Pergamon 1960), § 33
19. E. Fradkin, B.A. Huberman and S.H. Shenker, Phys. Rev. B 18, 4789 (1978)

20. S.M. Stishov, I.N. Makarenko, V.A. Ivanov and A.M. Nikolaenko, Phys. Lett. 45A, 18 (1973)
21. M. Lasocka, Phys. Lett. 51A, 137 (1975);
J.L. Tallon, Phys. Lett. 76A, 139 (1980)
22. M. Dixmier, J. Physique 39, 873 (1978)
23. S. Lefschetz, Introduction to Topology (Princeton 1949), ch.6/22

Universality in Size-Effects in 2D Percolation

J. Roussenq* and H. Ottavi**

*Département de Physique des Systèmes Désordonnés,
**Laboratoire d'Electronique
Université de Provence, F-13397 Marseille Cédex 4, France

1. Introduction

All experimental methods for percolation problems have the same disadvantage: obviously they are used on finite samples, whereas a mathematical problem is set on infinite lattices.

The results one can expect of such methods are related to a given size. This character can be used to study the variation of any property in terms of size and possibly one can extrapolate and infer the behaviour of the property for an infinite size.

Recently works of REYNOLDS [1] KIRKPATRICK [2] and ours [3] have used Monte-Carlo simulations to test the size dependence of the measured percolation threshold.

To this end we built many square samples of side n for each value of p, probability for an element to be active, and we measured the percentage of samples containing a continuous path of active elements between two opposite sides. This percentage gives an estimate of the conduction probability $F_n(p)$ between the considered sides.

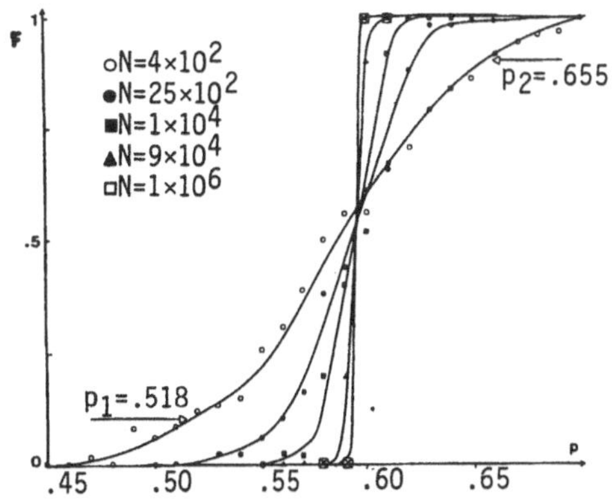

Fig.1 Gives the probability of conduction across a square network of $N=n^2$ elements versus the probability of site conduction.

Figure 1 shows the variation of $F_n(p)$ versus p for various size, in the case of the site problem on the square lattice. We can see that the passage from $F = 0$ to $F = 1$ is not abrupt, a transition zone exists whose the breadth decreases as the size increases.

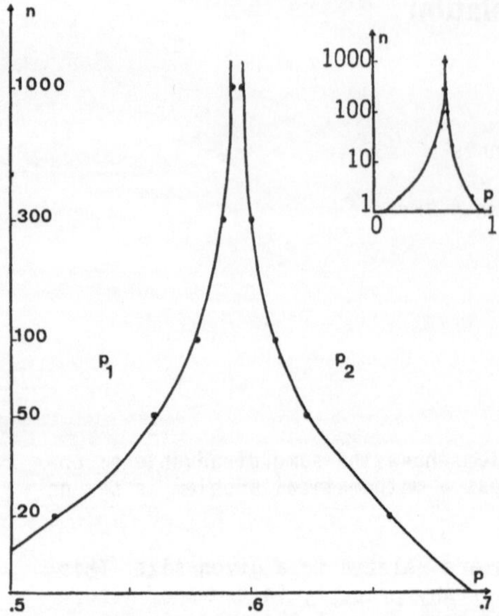

Fig.2 p_1 and p_2 are defined on figure 1 for $n = 20$. The inset gives the complete curve from n = 1 to 1 000 for a square network.

On figure 2 for guidance we plotted in semilog plot the values p_1 and p_2 of p for which $F(p_1) = 0.1$ and $F(p_2) = 0.9$ versus size of the samples.
Use of this result to give an estimate of the accuracy reached in determination of the percolation threshold using a given size of sample is evident.

But we can use this result in one another sense.

Consider the curve F of figure 3 ; symmetric with respect to the point (1/2,1/2) and corresponds to the bond problem of the square lattice for n = 3. We can do the piecewise-linearizing as in fig.3 drawing the tangent to the curve at the inflexion point, which intersects the p axis at p_m and F = 1 at p_M . The three segments of the jagged line F^* (joining 0,A (p_m , 0)), B (p_M, 1) and C (1,1) define three regions : two homogeneous OA and BC and a critical one AB.

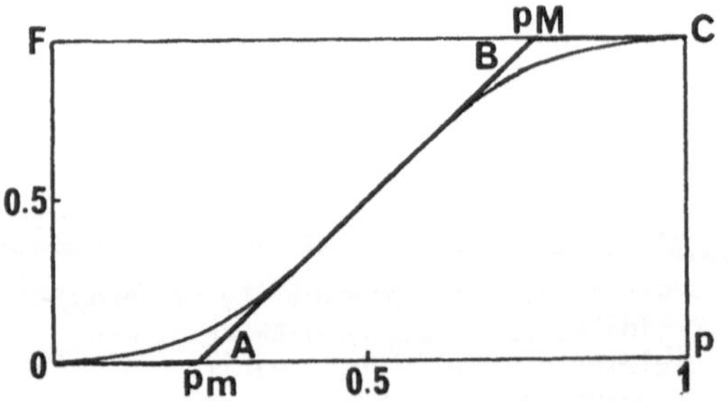

Fig.3 Piecewise-Linearizing for F_3

If we regard the correlation length as an average of the diameter of the clusters for $p < p_c$, we have

$$\xi \ll n \quad \text{for} \quad 0 < p < p_m \quad (\text{i.e. } F^*=0) \ .$$

For $p > p_c$, regarding ξ as an average of the diameter of the holes in the infinite cluster, we have

$\xi \ll n \quad$ for $\quad p_M < p < 1$ (i.e. $F^* = 1$)

For $p = p_m$ and $p = p_M$

$\xi \sim n$.

If λn is the maximum slope at the inflexion point

$$p_M - p_m = \frac{1}{\lambda n}$$

from

$$\xi(p_m) = \xi(p_M) \sim n$$

and

$$\xi(p) \propto \frac{1}{|p - p_c|^\nu} \quad \text{, we infer} \quad \lambda n \propto n1/\nu \quad .$$

We propose to test this hypothesis by mean of calculations and M.C. simulations on plane lattices in part 2, and study the shape of curves F in part 3.

2. Relation between Maximum Slope of F and Sample Size

The experiments have been made on three cases : bond problem on square lattice, site problem on square and triangular lattices.

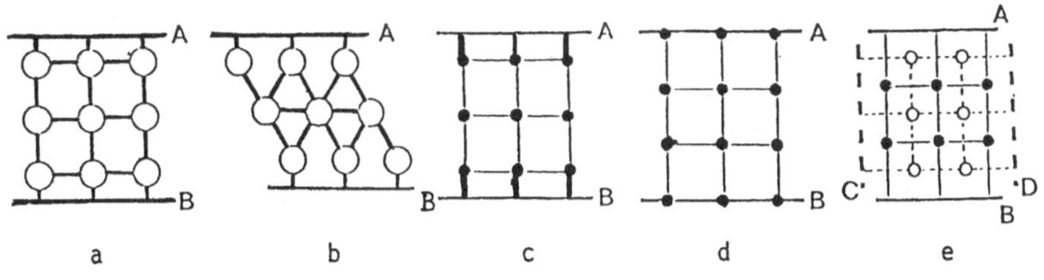

Fig.4a-e.

4a - Square lattice sample (site problem) of size n = 3. The bonds which are always conducting, are drawn in thick line ; the sites (which are conducting with the probability p) are in thin line . We study the probability F of occurence of a continuous conducting path between A and B.

4b- Triangular lattice sample (site problem), size n = 3. Identical comment as for figure 4a.

4d- Figure 4d represents a square lattice sample (bond problem), for n = 3. The sites are always conducting : they are drawn in thick line. The bonds, conducting at random, are in thin line. We choose this disposition in preference to that of figure 4c because it gives a center of symmetry to the curve F. Figure 4e is drawn to emphasize this property.

Figure 4 schematizes the networks we employed and the disposition of the sides between which we searched for a continuous conducting path.

We did not use the same methods for all the sizes of networks :

i- The direct exact calculation is possible for samples of size up to n = 3 or n = 4

ii- For larger samples this calculation is out of the question and we performed M.C. simulations for networks of size n = 10,20, 40, for which were made several thousands of trials for each value of p.

The experimental values of F obtained give an estimation of λn by an interpolation methods we describe in the following section. We have employed also the results of KIRKPATRICK [2] for networks of n = 64,200, 512 (bond problem on square lattice)

Results

We present the results we have obtained in log log plot on fig 5.

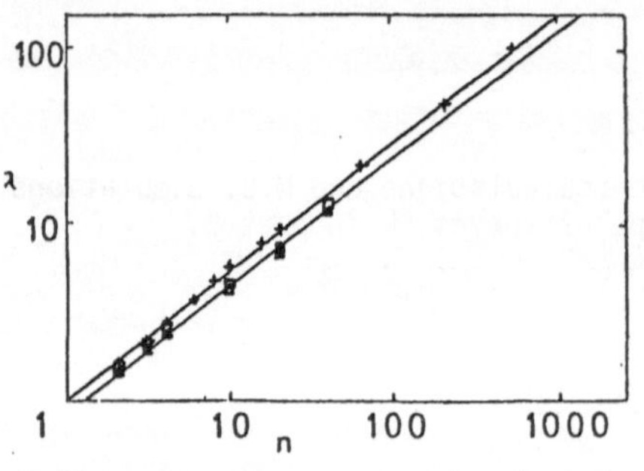

Fig.5a. Maximal slope versus size n, on a log log plot. The crosses represent the results for the square lattice (bond problem), the squares the results for the site problem on the square lattice and the triangles the results for the triangular lattice (site problem). We notice that the points are grouped on two distinct curves.

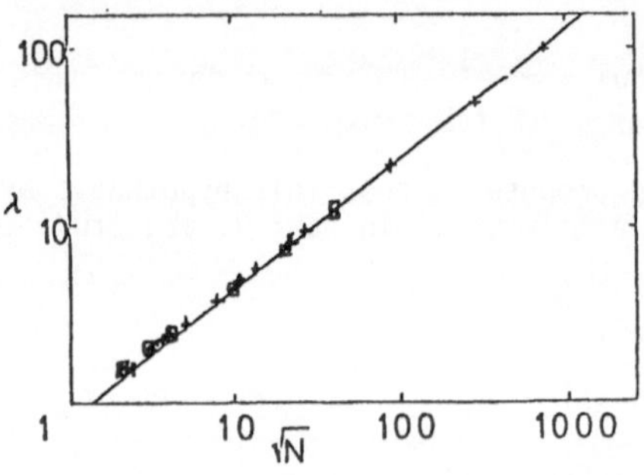

Fig.5b. The same points as in figure 5a are on an unique curve if we plot (1/2 log N) on the x-axis (N is the number of random elements in the sample).

One can see that those related to bond percolation on square lattice are on a curve and those related to both cases of site percolation are on a second curve (figure 5a).

These two curves are nearly straight lines and they are parallel. Their slopes, estimated for the largest values used for n are respectively :

$1/\nu b = 1/1.373$

$1/\nu s = 1/1.369$.

In both cases of site percolation, the number of elements N is such that $\sqrt{N} = n$ but for the case of bond percolation on square lattice we find :

$$N = n^2 + (n - 1)^2 .$$

If we plot on p-axis instead of log n, the quantity log $\sqrt{N}$, all the experimental points are on a single curve (fig.5b) .

This unique curve gives as exponent for large N :

νu = 1.385 (fitting all points of fig.5b) .

Accuracy

From the uncertainties on experimental datas (see appendix 1) one can establish that the uncertainty on ν u is less than 0.01 but it is not possible to specify how much the ν u found for N large but not infinite is different from the asymptotic value ν . Different estimates are obtained from the different models of extrapolation.

Other authors have used the same objects (finite size cells) in R.G.'s approach.

STANLEY et Al. [1] for square lattice, site problem found :

ν =1.354 ,

BERNASCONI [4] for bond problem on square lattice obtained

ν = 1.38 .

The differences between thes estimates are essentially due to the differences in the extrapolation methods.

3. Shape of the Curves F

3.1 Size dependence for a given lattice

We study the bond problem on square lattice. The exact calculation of the function F was done for n = 1,2 and 3 (see appendix 2)

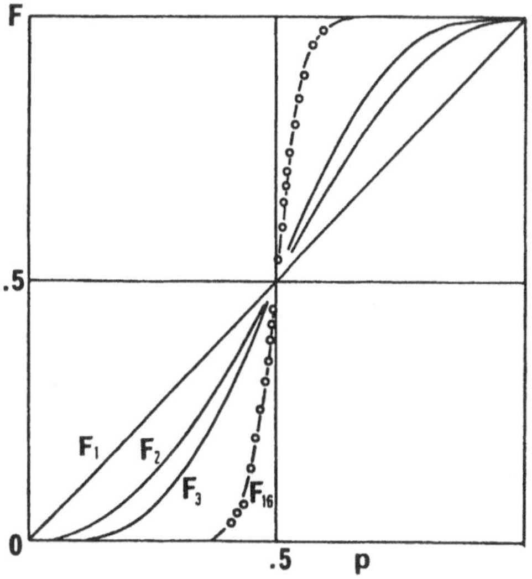

Fig.6 Bond problem on a square lattice. F_n is exactly calculated for n = 1,2 and 3. The points F_{16} represent the results obtained by Monte-Carlo simulations by KIRKPATRICK for n = 16

The curves are marked F_1, F_2, F_3, on figure 6 ; a center of symmetry exists of which the abscissa is $p = p_c = 1/2$. For size $n = 16$, we have used the data of M.C. simulations of ref [2] (points marked F_{16})

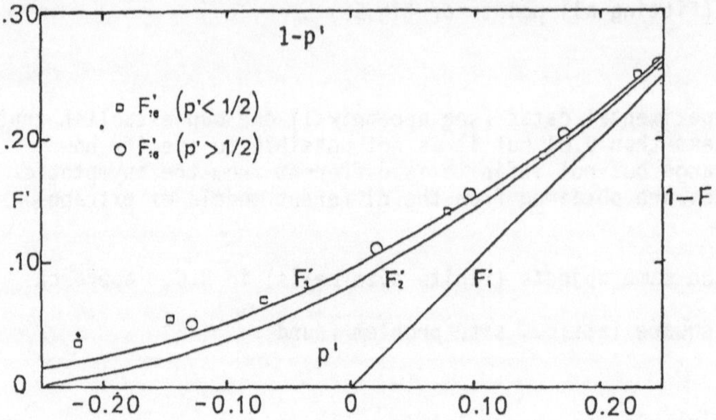

Fig.7 Each curve F_n' is deduced from the corresponding curve F_n of figure 6 by mean of a change of scale of the abscissas giving the value 1 to the slope at the inflexion point :

$p' = \lambda n(p - 0,5) + 0,5$ where λn is the maximum slope of F_n

On fig. 7 we have plotted the same curves after a change of scale on p-axis :

$$p' = \lambda n(p - 0,5) + 0,5$$

where λn is the slope of F_n at the inflexion point.

All the curves obtained by this method have a slope unity for $p' = 0,5$ (out of the boundaries of fig.7).

N.B. The data from ref[2] exhibit a slight dissymmetry probably due to the choice of disposition 4c (see fig.4) ; we laid to these points, in addition to an affinity a slight translation parallel to p-axis such that $p' = 0.5 + 7.5\ (p - 0.5 + 0.0039)$

One can see on fig.7 that the curves draw rapidly near a limit-shape as n increases. The datas for $n = 16$ are close to the curve F'_3 except for the points far away from the inflexion point.

3.2 Comparison between lattices

We compare the previous curves to those corresponding to a case with the same threshold and the same symmetry : site problem on triangular lattice. We have done the calculation for this case and $n = 1,2,3$ and 4 (appendix 2).

As in the previous case, we made an affinity parallel to p-axis, the point of abscissa 0.5 being invariant, and such that the slope at this point is unity. We foun that the curves obtained are intercalated perfectly well between those of fig.6

Particularly the curve for $n = 4$ in triangular lattice is practically identical with F'_3 ; discrepancies on F-axis are everywhere less than 0.001.

3.3 Comparison with a non-symmetric case

It is the case for the site problem in square lattice in which the inflexion point for $p = p_0$ is not a center of symmetry for the curve F ($F(p_0) \neq 1/2$).

However when the size increases, the dissymmetry decreases (appendix 3):

i- The value of F at the inflexion point draws near to 1/2 as it appears in appendix 3.
It is also interesting to point out that the point of inflexion becomes nearer p_c

ii- The shape of the curves draws nearer the shape of the symmetric ones of the previous cases.

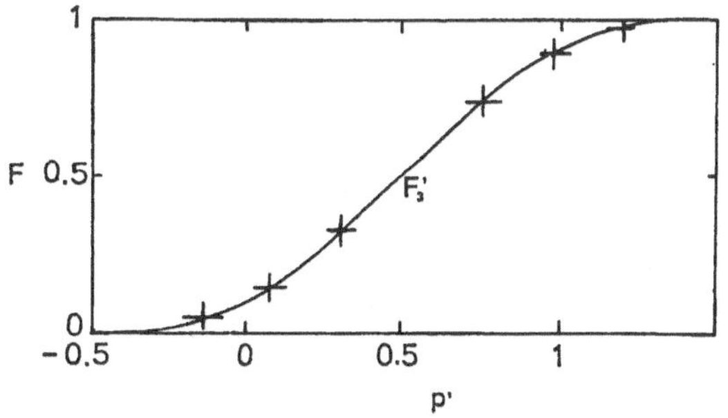

Fig.8 The continuous line is the curve F'_3 of fig.7. The crosses are calculated by Monte Carlo method for samples of size n = 10 of square lattice (site problem). For these points we make the change of scale defined by :

$$p' = 4.45\ (p - 0.584) + 0.5$$

This tendency to an unique shape is illustrated by fig.8 which represents :

- from one part the function F'_3
- from one another part the data obtained by M.C. method on samples of site problem on square lattice of size n = 10 .

The change of scale is $p' = 4.45\ (p - 0.584) + 1/2$ (parameters adjusted for the best fitting). The points all be on the curve F'_3.

To sum up, the curve F'_3 corresponding to bond problem on square lattice, seems to be convenient to represent the data related to any type of lattice, by the mean of an affinity and a translation parallel to the p-axis.

It is this curve we have used to interpolate our data.

4. Conclusion

This study points out two facts related to the notion of universality :

1/ The curve representing the probability of conduction of a large sized sample versus the bare probability p seems to have an unique shape.

2/ The maximum slope of this curve depends only on size of the sample.

These two results can be expressed by an unique relation first proposed by KIRKPATRICK :

$$F(p) \propto f[(p-pc)n^{1/\nu}]$$

where f appears to be independent of the lattice.

It seems possible to extend this point of view to the case of percolation with an external field. In the case of bond percolation on square lattice, let us suppose that there exists in addition to the bonds between sites, other bonds joining all the sites of the sample to the same "ghost site". If h is the probability of conduction of these latter bonds we postulate that the probability of conduction of the sample is asymptotically expressed by an homogeneous function.

$$C_{Tn}(p,h) \propto g[(p-pc)n^{1/\nu}, hn^{1/\nu_h}] \quad .$$

In compliance with the consideration expressed in ref [4] the knowledge of ν and ν_h is sufficient to determine all critical exponents.

Appendix 1.

Lattice	Origin	n	λ	σ
Square Bond Problem	Exact Calculation	1	1	0
		2	1.625	0
		3	2.217285	0
	Monte Carlo	4	2,69	0.03
		6	3.67	0.12
		8	4.76	0.16
		10	5.55	0.25
		16	7.60	0.08
		20	9.19	0.27
	Monte-Carlo (référence 1)	64	21.1	≈ 0.5
		200	48	≈ 1
		512	95	≈ 2
Triangular site problem	Exact Calculation	1	1	0
		2	1.5	0
		3	1.945313	0
		4	2.348633	0
	Monte Carlo	10	4.33	0.06
		20	7.18	0.10
		40	11.53	0.36
Square Site Problem	Exact Calculation	1	1	0
		2	1.539601	0
		3	1.996275	0
		4	2.416614	0
	Monte Carlo	10	4.45	0.10
		20	7.21	0.24
		40	12.72	0.26

Appendix 2.1

n	F - square lattice - bond problem
1	p
2	$2p^2 + 2p^3 - 5p^4 + 2p^5$
3	$3p^3 + 8p^4 + 2p^5 - 37p^6 - 10p^7 + 39p^8 + 149p^9 - 352p^{10} + 298p^{11}$ $- 117p^{12} + 18p^{13}$

Appendix 2.2

n	F - Triangular lattice - Site problem
1	p
2	$3p^2 - 2p^3$
3	$8p^3 - 6p^4 - 6p^5 + 0p^6 + 12p^7 - 9p^8 + 2p^9$
4	$20p^4 - 12p^5 - 30p^6 - 20p^7 + 45p^8 + 100p^9 - 66p^{10} - 300p^{11} + 520p^{12}$ $- 360p^{13} + 120p^{14} - 16p^{15}$

Appendix 2.3

n	F - Square lattice - site problem
1	p
2	$2p^2 + 0p^3 - p^4$
3	$3p^3 + 4p^4 - 6p^5 - 9p^6 + 14p^7 - 6p^8 + p^9$
4	$4p^4 + 12p^5 - 6p^6 - 28p^7 - 22p^8 + 48p^9 + 66p^{10} - 108p^{11} + 10p^{12}$ $+ 44p^{13} - 20p^{14} + 0p^{15} + p^{16}$

Appendix 3.

Shape of curve F for non symmetric case : inflexion point is not center of symmetry . For a given size let us consider Taylor's development near this point :

$$F = a_0 + a_1(p-p_0) + a_2(p-p_0)^2 + a_3(p-p_0)^3 + \ldots \quad . \qquad (1)$$

The non symmetry appears in two features:

- a_0 is different from 1/2
- The others even terms are not all zero (a_2=0 since p_0 is the abscissa of the inflexion point)

When size increases, dissymmetry decreases :

1- Table 3a shows that $F(p_0)$ becomes nearer 1/2 (and $p_0 \rightarrow p_c \simeq 0,593$)

2- The contribution of even termes decreases with : $a_1 = \lambda$

$$a_k = C_k \lambda^k \quad .$$

Eq (1) becomes

$$F - a_0 = \lambda(p-p_0) + c_3\lambda^3(p-p_0)^3 + c_4\lambda^4(p-p_0)^4 + \dots \quad . \qquad (2)$$

Two identical series C_k correspond to two identical curves.
In table 3b we give the numerical values we have calculated for C_3; C_4, C_5 in the Three cases of percolation.
For the site problem on square lattice .We verified by another way that $|C_4|$ decreases when n increases , denoting a tendence to symmetry

Appendix 3a.

Size	p_0	$a_0 = F(p_0)$
2	0,577350	0,555556
3	0,580302	0,541875
4	0,584400	0,535425

Appendix 3b.

Lattice	C_3	C_4	C_5
Square lattice Site problem - n=4	-0,895191	-0,142897	0,534775
Triang. Lattice Site problem - n=4	-0,890989	0	0,560831
Sq. Lattice - Bond problem - n=3 (F_3')	-0,893342	0	0.571291

References

1. P.J. REYNOLDS, H.E. STANLEY and W. KLEIN -J. Phys A 11, N°8, p L199-207 (1978) et nombreux autres articles dont un paru dans Phys.Rev. B, le 1.02. 1980

2. Models of Disordered materials (Les Houches été 1978) par S. KIRKPATRICK IBM Res. Center, Yorktown heights, N.Y. 10598

3. J.ROUSSENQ,J. CLERC, G.GIRAUD,E. GUYON and H. OTTAVI - J.de Phys. Lettres 37, p L99-101 (1976)

4. J. BERNASCONI, Phys. Rev. B 18, N°5, p 2185-2191 (1978)

2.2 Use of Renormalisation Techniques

The Phenomenological Renormalization Method

B. Derrida

Service de Physique Théorique, CEN Saclay, B.P. Nr. 2
F-91190 Gif sur Yvette, France

J. Vannimenus

Laboratoire de Physique de l'Ecole Normale Supérieure, 24, Rue Lhomond
F-75231 Paris Cédex 05, France

1. Introduction

One aim of the study of critical phenomena is the calculation of the exponents which describe the singularities of the thermodynamic functions at a second order phase transition point. Most of the models of statistical mechanics are not solvable exactly, therefore these exponents are only known approximatively. The three main approaches which allow to calculate the critical exponents are the series expansions, the Monte Carlo methods and the renormalization group theories. The phenomenological renormalization method which was introduced by NIGHTINGALE [1] is a real space renormalization method. The philosophy of the method is to use the fact that one can calculate exactly the thermodynamic properties of one dimensional systems and then from this information obtain the critical properties of systems in higher dimension. Its main advantages are :

1) the method can be used for a large class of models,
2) it gives satisfactory results which can be improved systematically,
3) only one parameter is renormalized.

2. The Method

Suppose that one wants to calculate the critical temperature and the critical exponents of a two dimensional system (for example the Ising model or the Potts model). For simplicity the only parameter in the model is the temperature T. Using the transfer matrix method (see the appendix), one can obtain exactly the thermodynamic quantities for an infinite strip of width n (see figure 1)

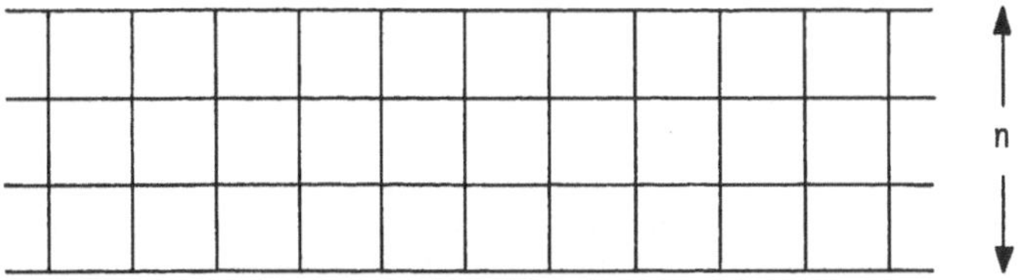

Fig. 1 An infinite strip of width n

The correlation length $\xi_n(T)$ along the strip is given by :

$$\frac{1}{\xi_n(T)} = \log \frac{\lambda_1(T)}{\lambda_2(T)} , \qquad (1)$$

where λ_1 and λ_2 are the two largest eigenvalues of the transfer matrix.

From the knowledge of the correlation lengths $\xi_n(T)$ and $\xi_m(T)$ of two strips of widths n and m, one can write the phenomenological renormalization equation :

$$\frac{\xi_n(T)}{\xi_m(T')} = \frac{n}{m} . \tag{2}$$

This equation establishes a correspondence between the two strips : the strip of width m at temperature T' is related by a scaling transformation to the strip of width n at temperature T. This transformation is a contraction of ratio n/m of the width of the strip and so the correlation length which is the characteristic length along the strip has to be contracted in the same ratio.

3. Validity of the Method

Equation (2) gives a relation between T' and T

$$T' = R_{n,m}(T) . \tag{3}$$

If the renormalization $R_{n,m}$ depended only on the ratio n/m, then the relation (3) would be valid for the two dimensional system ($n \to \infty$ and $m \to \infty$ with n/m fixed).

In fact $R_{n,m}$ does not depend only on the ratio n/m, so (3) is only an approximate renormalization equation for the two dimensional system. As we discuss it later in the case of the Ising model, the best results are obtained when n is large and m=n-1.

4. Critical Point and Critical Exponents [2]

From the renormalization equation (2), it is straightforward to calculate the critical temperature T_c : T_c is the fixed point of the recursion formula (2)

$$\frac{\xi_n(T_c)}{\xi_m(T_c)} = \frac{n}{m} . \tag{4}$$

Assuming again that the relation (2) between T and T' depends only on the ratio n/m, one has

$$\frac{\xi_n(T)}{\xi_m(T')} = \frac{n}{m} = \frac{\xi(T)}{\xi(T')} , \tag{5}$$

where $\xi(T)$ is the correlation length of the two dimensional system. Note that for $T > T_c$, $\xi_n(T) \to \xi(T)$ for large n and relation (5) is verified asymptotically.

Now it is easy to calculate the exponent ν of the two dimensional correlation length

$$\xi(T) \sim |T-T_c|^{-\nu} .$$

From (5) one has

$$\nu = \frac{1}{\dfrac{\log\left(\dfrac{d\zeta_n}{dT} \bigg/ \dfrac{d\zeta_m}{dT}\right)}{\log(n/m)} - 1} \tag{6}$$

where the derivatives are calculated at the critical point T_c solution of (4).

Suppose we want to calculate another critical exponent ω which describes the critical behaviour of a quantity P(T) of the two dimensional system :

$$P(T) \sim (T - T_c)^{\omega} \quad . \tag{7}$$

From the transfer matrix method, one can calculate exactly $P_n(T)$ the value of this quantity for a strip of width n.

Assuming again that $\frac{P_n(T)}{P_m(T')}$ depends only on the ratio n/m, one has

$$\frac{P_n(T)}{P_m(T')} = \frac{P(T)}{P(T')} \; .$$

It follows that

$$\frac{P_n(T_c)}{P_m(T_c)} \sim \left(\frac{n}{m}\right)^{-\omega/\nu} \quad . \tag{8}$$

So using the renormalization (2) for a given choice of (n,m), we can find approximate values of T_c, ν and any other exponent ω from equations (4),(6) and (8).

5. The Ising Model

In order to estimate the accuracy of the method, it is interesting to apply it to the two dimensional Ising model [1].

For the two dimensional Ising model, the Hamiltonian is

$$\mathcal{H} = - J \sum_{<ij>} \sigma_i \, \sigma_j \tag{9}$$

where $\sigma_i = \pm 1$ and the sum is performed over all the pairs of nearest neighbours on the square lattice. Call $K = J/T$. The Ising model has been solved exactly and the exact values of K_c and ν are :

$$K_c = \frac{1}{2} \log\,(1+\sqrt{2})$$

$$\nu = 1 \; .$$

The expression of the correlation length ξ_n for strips of width n with periodic boundary conditions is known exactly [1] :

$$\frac{1}{\xi_n(K)} = \frac{1}{2}\,(\gamma_1+\gamma_3+\ldots+\gamma_{2n-1}) - \frac{1}{2}\,(\gamma_o+\gamma_2+\ldots+\gamma_{2n-2}) \quad , \tag{10}$$

where the γ are given by

$$\cosh \gamma_r = \frac{\cosh^2 2K}{\sinh 2K} - \cos \frac{r\pi}{n} \quad . \tag{11}$$

Using these expressions one can expand the correlation length around the critical point K_c for large n.

$$\frac{1}{\xi_n} = \frac{\pi}{4n} + \frac{\pi^3}{96n^3} + \ldots - \frac{\gamma_o}{2} + \gamma_o^2 \left[\frac{n}{2\pi} \log 2 + \ldots \right] + O(\gamma_o^3) \tag{12}$$

where γ_o is defined by (11) and behaves like

$$\gamma_o \simeq 4(K-K_c) \quad .$$

Now we can use the phenomenological method to calculate the approximate critical point $K_c(n)$ and the exponent $\nu(n)$ in the large n limit and with $m = n/\lambda$. The results are :

$$K_c(n) = \frac{1}{2} \log(1+\sqrt{2}) - \frac{\pi^3}{192} \frac{\lambda(\lambda+1)}{n^3} + \ldots$$
$$\nu(n) = 1 - \frac{\pi^2 \log 2}{24\, n^2} \frac{\lambda^2-1}{\log\lambda} + \ldots \tag{13}$$

The convergence is rather rapid and is optimal for λ as close to 1 as possible. Therefore the best choice for m is m=n-1. This choice is physically not surprising because it minimizes the size effects.

6. Other Models

This phenomenological renormalization method has been successfully used in various situations : several generalized Ising models [2], the spin s Ising model [3], the antiferromagnetic Ising model in a magnetic field [4], the q state Potts model [5].

We have used this phenomenological renormalization to study the site and the bond percolation problem in dimension 2. We give here the results obtained for the site percolation problem [6]. The correlation lengths have been calculated for strips of width n with periodic boundary conditions as functions of p, the probability for a site to be occupied. Table 1 contains the values of the percolation threshold p_c and the exponent ν we have found using the phenomenological renormalization method for several values of n and m. The size s_n of the transfer matrix for a strip of width n is given to indicate the amount of work involved.

n	m	p_c	ν	s_n
2	1	.734	1.56	2
3	2	.582	1.49	3
4	3	.591	1.47	5
5	4	.589	1.41	7
Accepted values from other methods		.5935±.001	1.34 ± .02	

Table 1

This example illustrates the interest of the phenomenological renormalization method as satisfactory results are obtained with a reasonable amount of calculations. Moreover the method can be improved by using strips of larger width. Its main advantage is that it can be simply used for a large class of models. It allows also studying models in dimension 3, though the size of the transfer matrix increases more rapidly in this case.

Appendix

In this appendix we explain how to compute the correlation length of two examples : the Ising chain and a strip of width 2 for the site percolation problem.

The Ising Chain

Consider an Ising chain of N spins. The Hamiltonian is given by

$$\mathcal{H} = -J \sum_{i=1}^{N-1} \sigma_i \sigma_{i+1} \ .$$

The partition function Z_N of this system is

$$Z_N = \sum_{\sigma_i = \pm 1} \exp\left(K \sum_i \sigma_i \sigma_{i+1} \right) \ . \tag{A1}$$

If one wants to find a recursion formula which relates Z_N to Z_{N+1}, one has to cut Z_N into two parts : $Z_N(+)$ and $Z_N(-)$. $Z_N(\sigma)$ is the part of the sum (A1) where the spin σ_N has the value σ

$$Z_N = Z_N(+) + Z_N(-) \ .$$

The transfer matrix is the matrix which relates the $Z_{N+1}(\sigma)$ to the $Z_N(\sigma)$. In the case of the Ising chain one has :

$$\begin{pmatrix} Z_{N+1}(+) \\ Z_{N+1}(-) \end{pmatrix} = \begin{pmatrix} e^K & e^{-K} \\ e^{-K} & e^K \end{pmatrix} \begin{pmatrix} Z_N(+) \\ Z_N(-) \end{pmatrix} \ .$$

For a more complicated 1 dimensional model, the size of the transfer matrix is the number of states of the spins in the Nth column.

The partition function Z_N behaves like

$$Z_N \sim \lambda_1^N \left(\alpha + \beta \left(\frac{\lambda_2}{\lambda_1}\right)^N + \dots\right)$$

where λ_1 and λ_2 are the two largest eigenvalues of the transfer matrix. So the free energy is given by λ_1 while λ_2/λ_1 gives the characteristic length along which the influence of the boundary conditions of column 1 is damped.

A Strip of Width 2 for the Site Percolation Problem

p is the probability for a site to be occupied.

call :
- $A_N(p)$ the probability that the two sites of the Nth column are connected to the 1st column.
- $B_N(p)$ the probability that only one site of the Nth column is connected to the 1st column.
- $C_N(p)$ the probability that no site of the Nth column is connected to the 1st column.

It is easy to relate these quantities at column N and at column N+1.

$$\begin{pmatrix} A_{N+1} \\ B_{N+1} \\ C_{N+1} \end{pmatrix} = \begin{pmatrix} p^2 & p^2 & 0 \\ 2p(1-p) & p(1-p) & 0 \\ (1-p)^2 & (1-p) & 1 \end{pmatrix} \begin{pmatrix} A_N \\ B_N \\ C_N \end{pmatrix} .$$

The largest eigenvalue is 1 (conservation of the total probability). One can verify that

$$A_N \text{ or } B_N \sim \lambda^N$$

where λ is the largest root of

$$\lambda^2 - \lambda p - p^3(1-p) = 0 .$$

As the probability for the Nth column to be connected to the 1st column behaves like λ^N, the correlation length is given by

$$e^{-1/\xi} = \lambda .$$

References

1. M.P. Nightingale, Physica 83A, 561 (1976)
2. M.P. Nightingale, Proceedings of the Koninklijke Nederlandse Akademie van Wetenschappen, B82 (3), 235 (1979)
3. L. Sneddon, J. Phys. C11, 2823 (1978)
4. L. Sneddon, J. Phys. C12, 3051 (1979)
5. M.P. Nightingale, H.W.J. Blöte, Preprint 1980.
6. B. Derrida, J. Vannimenus, Preprint (1980), CEA DPh-T 62/80.

Computation of the Yang-Lee Edge Singularity in Ising Models

P. Moussa

Département de Physique Théorique, Centre d'Etudes Nucléaires de Saclay
F-91190 Gif sur Yvette, France

1. Introduction

In 1952, Yang and Lee [1] established the link between the occurence of phase transition, and the location of zeroes of the partition function in the complex plane of a suitable activity variable. Lee and Yang [2] analyzed the location of zeroes for the lattice gaz as well as for the ferromagnetic spin one half Ising model. In the thermodynamic (large volume) limit these zeroes will become dense and form a cut in the complex plane. When the temperature is larger than the critical one, this cut ends up into a pair of singularities, which move with the temperature. The critical singularity is generated by the meeting of these two singularities (and possibly others). Nothing is known exactly about the nature of the singularities for the well-known Ising model in more than one dimension. This lack of knowledge indicates that a global understanding of the critical point in the two variables, temperature and magnetic field (or order parameter) has not yet been achieved. Our purpose is to recall the various attempts which have been made in this direction, and also to show why the problem is particularly hard. Although we shall see that these singularities occur for a pure imaginary field and therefore look highly unphysical, we think that the understanding of these singularities remains a challenge for physicists working in statistical mechanics.

2. The Lee-Yang Representation in the Ferromagnetic Ising Model

We shall consider the ferromagnetic model in dimension d. The partition function for a finite subset of N spins sitting on a regular lattice is given by

$$Z_N = \sum_{\{\sigma\}} \exp - \beta \left[-J \sum_{n.n} \sigma_i \sigma_j - h \sum_i \sigma_i \right] \tag{1}$$

where σ_i takes the values $\pm$ 1 for i = 1,2,...N, J is real and positive for ferromagnetic systems, and the coupling sum runs over pairs of nearest neighbours. h is the magnetic field. All Boltzmann factors are added up for each possible values of the σ_i's. We shall use the following variables :

$$x = e^{-2\beta J} \quad , \quad z = e^{-2\beta h} \quad , \quad \beta = \frac{1}{kT} \; . \tag{2}$$

x is a temperature variable which runs from zero to one when T runs from zero to infinity.

Lee an Yang [2] have proved that for fixed T (real and positive) all zeroes of Z_N sit on the unit circle $|z|$ = 1, in the complex plane of the variable z, that is for purely imaginary field. Apart from an irrelevant factor, Z_N is a polynomial in z of degree N and therefore can be factorized as

$$Z_N = \prod_{i=1}^{N} (z - z_i(x)) \ . \tag{3}$$

The free energy per spin is defined as

$$F(x,z) = \lim_{N\to\infty} \left(- \frac{1}{\beta N} \ln Z_N \right) . \tag{4}$$

The factorization of Z_N allows one to write log Z_N as a sum and F becomes an integral in the limit $N \to \infty$. Explicit expressions for F and the magnetization are then given by the Lee and Yang representation :

$$\beta F = -\beta h - \frac{c\beta J}{2} - \int_{\theta_o(x)}^{\pi} \ln(z^2 - 2z \cos\theta + 1)\, g(\theta,x)\, d\theta \tag{5}$$

$$M = - \frac{\partial F}{\partial h} = 2(1-z^2) \int_{\theta_o(x)}^{\pi} \frac{g(\theta,x)}{1 - 2z\cos\theta + z^2}\, d\theta \ , \tag{6}$$

θ describes the angle on the circle $|z| = 1$, $g(\theta,x)$ is the limit density of zeroes, that is a positive measure. For $T \leq T_c$, $\theta_o = 0$, and when T runs from T_c to ∞, θ_o increases from 0 to π. c is the coordination number of the model.

In the high temperature phase, the zeroes do not cover the whole circle. The critical point is obtained when the so called Yang-Lee edge singularitie at $z = e^{\pm i\theta_o}$ move on the circle and meet at the point z = 1.

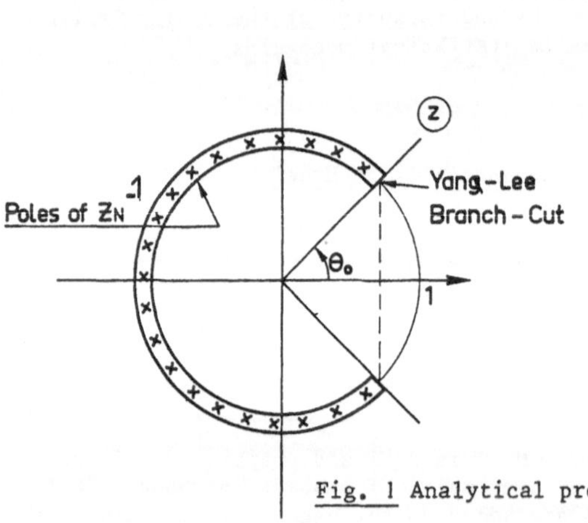

Fig. 1 Analytical properties of the magnetization for T > Tc

The analytical properties of the magnetization as function of z are displayed in Fig.1 for $T > T_c$. We have the following symmetry under reversal of the magnetic field :

$$M(z) = - M(\tfrac{1}{z}) \ . \tag{7}$$

Since M is analytic at z = 1, M vanishes when h is equal to zero and we have no spontaneous magnetization for $T > T_c$.

The situation is different for a purely imaginary field :

$$h = -\frac{i\theta}{2\beta} \tag{8}$$

A discontinuity occurs for $\theta = \theta_o$ in the real part of the magnetization. This can be seen through the following formula, which is easily deduced from (6) :

$$M = 2\left(\frac{1-z}{1+z}\right)\int_{\theta_o}^{\pi} \frac{g(\theta,x)\ d\theta}{1-(4z/(1+z)^2)\cos^2\theta/2} \tag{9}$$

$$\frac{M}{\sqrt{1-v}} = 2\int_{\theta_o}^{\pi} \frac{g(\theta,x)\ d\theta}{1-v\cdot\cos^2\theta/2} \tag{10}$$

where $v = \frac{4z}{(1+z)^2} = 1 - \tanh^2(\beta h)$. v is real and greater than one when h is pure-imaginary. Since formula (9) is a representation of M as Stieltjes function [3], it is suitable for analysis of the Yang-Lee singularity, which is the nearest one in the v plane for $T > T_c$. The discontinuity of the function $\frac{M}{\sqrt{1-v}} = \frac{M}{\tanh(\beta h)}$ as a function of v across the real axis is therefore purely imaginary for $v > v_o = \frac{1}{\cos^2\theta/2}$, and equal to :

$$\left[\frac{M}{\sqrt{1-v}}\right]_{v=v+i\varepsilon} - \left[\frac{M}{\sqrt{1-v}}\right]_{v=v-i\varepsilon} = 2\pi i g(\theta,x) \tag{11}$$

in which we set $v = \frac{1}{\cos^2\theta/2}$.

The discontinuity of M across the circle in the plane of the variable z is therefore purely real.

Finally we shall define the Yang Lee edge singularity index σ as :

$$g(\theta,x) \simeq A(\theta-\theta_o)^\sigma \text{ for } \theta \text{ close to } \theta_o.$$

For the magnetization in a purely imaginary field, $h = ih^I$

$$m(h) \sim (h^I - h_{YL})^\sigma \quad \text{for } h^I \text{ close to } h_{YL}$$

with

$$h_{YL} = \frac{\theta o}{2\beta}.$$

However for $T = T_c$, $h_{YL} = 0$, and we have a cross-over effect to the usual critical behaviour $M \sim h^{1/\delta}$.

In terms of the density of zeroes, one has at the critical temperature

$$g(\theta,x_c) \sim \theta^{(1/\delta-1)} \text{ for } \theta \text{ small.}$$

The additional (-1) term come from the extra factor (1-z)/(1+z) in (9). We mention here for completeness that the behaviour of θ_o with T is governed by a different critical index :

$$\theta_o(T) \simeq (T - T_c)^{\Delta}$$

where Δ is the usual gap index [4].

Before going into the numerical evaluation of the σ index, we want to mention that the Lee-Yang circle theorem has been generalized to various models : ferromagnetic models for spin greater than one half [5], including many-spin interaction [6]. A very elegant proof which allows generalization to complex temperature plane has been given by Ruelle [7,8]. The relation with the classical Griffiths inequalities has been analyzed by Asano [9]. The generalization to Heisenberg models can be found in the works of Dunlop and Newman [10,11]. A good review has been given recently by D.A. Kurtze [12].

3. The Yang-Lee Singularity Index

The exact value of σ is known only for few models : $\sigma = 1/2$ for mean field theory [13], for the Bethe lattice [3], for the spherical model [14], and for the one dimensional hierarchical model [15]. $\sigma = -1/2$ for the one dimensional Ising model [2]. In other words exact results are known only when the Yang Lee edge singularity appears to be of the square root type.

Numerical investigation of the Ising model for dimension higher than one have been initiated by Kortman and Griffiths [13], and pursued by Kurtze and Fisher [16]. We shall now discuss their results. Kortman and Griffiths applied Neville table analysis to series expansion of the magnetization in powers of the variable tanh (βh). They gave the first evaluation of σ for the two and three-dimensional Ising model. However they observed that their results depend strongly on the value of the temperature. Therefore their estimate is very rough. Nevertheless they state that σ is negative and small ($\sigma \sim -0.125$) in the two-dimensional Ising model and positive and small ($\sigma \sim +0.125$) in the three-dimensional one. They also gave the first construction of the measure g(θ) through its Fourier coefficients.

Kurtze and Fisher [16] argued that the best value must be obtained in the high temperature limit, namely far away from the cross-over occuring at the critical temperature. They show that the high temperature limit is in itself interesting because of its relation with the monomer-dimer problem. This connection was also noticed by Baker and Moussa [17].Kurtze and Fisher [16] considered various refined techniques for extracting the index σ from series expansion at high temperature and they quote $\sigma = -0.163 \pm 0.003$ in two dimensions, and $\sigma = 0.086 \pm 0.015$ in three dimensions. These number are in agreement with former given by Gaunt [18] concerning the monomer dimer problem.

In order to clarify the situation, I have considered the variation with the temperature either in the Bethe lattice model with coordination number c = 4, or in the corresponding two dimensional Ising model. Evaluation of σ has been performed using Pade approximants on the logarithmic derivative of the series expansion [3] in powers of the variable v defined above in equation (10). The so-called D-log Pade method [19], first introduced by Baker and Gammel, converts a branch-cut singularity $(x-x_c)^{\sigma}$ with $\sigma < 0$, into a polar singularity $\frac{\sigma}{x-x_c}$. When $0 < \sigma < 1$, we first derive the function in order to get a singularity $(x-x_c)^{\sigma-1}$ to which the method applies. Then, we use Pade approximants to obtain from series, approximate values for the poles x_c and the residue σ.

The results of our analysis are displayed in figure 2 :

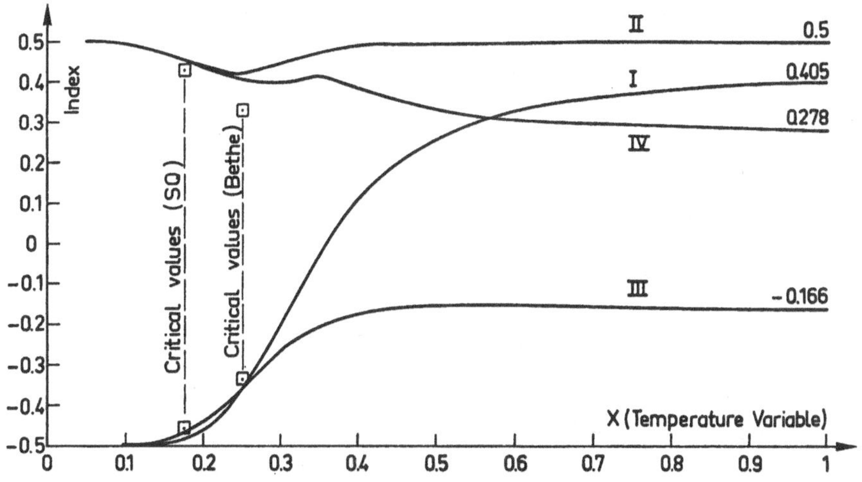

Fig. 2 Lee Yang singularity index calculated from magnetization series, I Bethe lattice, II Bethe lattice (Inverted series), III Square lattice,, IV Square lattice (Inverted series)

Curve I shows the variation of σ with the temperature for the Bethe lattice. Notice that for $x < x_c$, the expected index is $(-\frac{1}{2})$ according to the existence of a spontaneous magnetization, which generates a singularity of the $M/\sqrt{1-v}$ type in the considered series. The high temperature value is 0.405 which is far from the exact 0.5 value. This discrepancy comes from the existence of a hidden pole just behind the Yang-Lee singularity which we try to analyse [17]. This extrapole is exactly located at the branch cut in the c = 2 case, converting the index from $+\frac{1}{2}$ to $-\frac{1}{2}$, which is the result in the one-dimensional Ising model. In the Bethe lattice (c=4) case, we were able to get rid of the difficulty in analysing the inverse series, that is $\frac{\tan\beta h}{M}$. Result are given in curve II. Except in the vicinity of the cross-over point, we see a perfect agreement with the exact 0.5 value. More precisely for $x > 0.4$, the discrepancy is less than 2.10^{-3}.

Results for the Ising model are given in curve III for $\frac{M}{\tan\beta h}$, and in curve IV for the inverted series $\frac{\tan\beta h}{M}$. In curve III we get at high temperature agreement with Kurtze and Fisher, however the inverted series gives a value of $\sigma \simeq -0.278$ at $T \to \infty$. The discrepancy between - 0.166 and - 0.278 shows that the situation is not clarified and that we must keep open the possibility that a hidden singularity is responsible for the discrepancy. In my opinion any intermediate value of σ in between remains believable. Magic-numbers lovers can still dream of a value like $-\frac{1}{4}$. The fact that the two series give results of opposite signs is at least a proof that σ must be negative.

Although we are not able to give a precise value we think that the results remain compatible with the renormalization group analysis performed by Fisher [20]. Starting from the observation that the symmetry under reversal of the magnetic field is broken, Fisher considers a universal Hamiltonien similar to the one given by Landau, but with an additional cubic term. Therefore the critical dimension should be 6. That is for $d > d_c = 6$, $\sigma = 0.5$, and an expansion in powers of (6-d) is possible for $d < 6$. But the number of terms available is small and extrapolation from d = 6 to d = 2 cannot be accurate. The curve given by Fisher [20] surely allows σ to take values up to - 0.3 without destroying the analysis. Furthermore, evaluation using direct space renormalization group analysis [21] have been obtained for the one-dimensional quantum Ising chain with a transverse field which is known to be a particular case of the classical two-dimensional case [22]. Uzelac et al [23] give a value of $\sigma \sim -0.27$. Further calculations using this method are still in progress. For more details see [24].

4. Conclusion

We have seen that the evaluation of the Lee Yang singularity index is not a simple problem, and we still need a better knowledge of the nature of this singularity. An exact result would be a great progress for instance in the monomer dimer problem. Some attempts have been made [25] which relate this problem to a somewhat unusual mathematical moment problem with integer moments. The Yang Lee singularity problem has also provided some surprises will the $\sigma = 0.5$ results in the spherical [14] and hierarchical [15] models. In these two models renormalization group analysis would at first sight give non classical results ($\sigma \neq 1/2$) at least for some values of the parameters. The exact reason for the non-applicabily of the renormalization group analysis is not known but might be thought to be related either to the interferences due to the complex field or to the lack of translational invariance for finite systems (no periodic boundaries are possible). We expect that the Lee Yang singularity puzzle will also give better understanding of the renormalization group "approach" itself.

We thank G. Baker, B. Derrida, R. Jullien, M. Fisher, D. Kurtze, P. Pfeuty and K. Uzelac for valuable discussion and communication of their results prior to publication.

References

1. C.N. Yang and T.D. Lee, Phys. Rev., 87, 404 (1952).
2. T.D. Lee and C.N. Yang, Phys. Rev., 87, 410 (1952).
3. D. Bessis, J.M. Drouffe and P. Moussa, J. Phys.A : Math. Gen., 9, 2105 (1976).
4. M.E. Fisher, Rep. Prog. Phys., 30, 615 (1967).
5. R.B. Griffiths, J. Math. Phys., 10, 1559 (1969).
6. M. Suzuki and M.E. Fisher, J. Math. Phys., 12, 235 (1971).
7. D. Ruelle, Phys. Rev. Letters, 26, 303 (1971).
8. D. Ruelle, Commun. Math. Phys., 31, 265 (1973).
9. T. Asano, Phys. Rev. Letters, 24, 1409 (1970).
10. F. Dunlop and C.M. Newman, Commun. Math. Phys., 44, 223 (1975).
11. C.M. Newman, J. Stat. Phys., 15, 399 (1976).
12. D.A. Kurtze, Ph D. Thesis, Cornell University (1980), MSC report N°4184 (1979).
13. P.J. Kortman and R.B. Griffiths, Phys. Rev. Lett., 27, 1439 (1971).
14. D.A. Kurtze and M.E. Fisher, J. Stat. Phys., 19, 205 (1978).
15. G.A. Baker, M.E. Fisher and P. Moussa, Phys. Rev. Lett., 42, 615 (1979).
16. D.A. Kurtze and M.E. Fisher, Phys. Rev.B, 20, 2785 (1979).
17. G.A. Baker and P. Moussa, J. Appl. Phys., 49, 1360 (1978).
18. D.S. Gaunt, Phys. Rev., 179, 174 (1969).

19. D.S. Gaunt and A.J. Guttmann, in "Phase Transitions and Critical Phenomena", Academic Press, New York (1974), Vol.3, p.181.

20. M.E. Fisher, Phys. Rev. Lett., 40, 1610 (1978).

21. K. Uzelac, P. Pfeuty and R. Jullien, Phys. Rev. Lett., 43, 805 (1979).

22. M. Suzuki, Prog. Theor. Phys. 56, 1454 (1976).

23. K. Uzelac, R. Jullien and P. Pfeuty, Phys. Rev.B, Vol.22 (1980) in press.

24. K. Uzelac et al, Communication to this conference.

25. M.F. Barnsley, D. Bessis and P. Moussa, J. Math. Phys., 20, 535 (1979).

Real-Space Renormalization-Group Method for Quantum Systems: Application to Quantum Frustration in Two Dimensions

K.A. Penson*, R. Jullien, P. Pfeuty, and K. Uzelac

Laboratoire de Physique des Solides, Bâtiment 510, Université Paris Sud
Centre d'Orsay
F-91405 Orsay Cédex, France

A real-space renormalization-group method well suited for studying phase transitions at T = 0 in quantum systems is presented and applied to a generalized spin 1/2 planar hamiltonian on the triangular lattice, which includes ferromagnetic and antiferromagnetic XY models as limiting cases. This provides a description of the effects of frustration in quantum systems.

While the theory of phase transitions has been mainly developed for classical spin systems, the study of quantum systems has only been recently an object of increasing theoretical interest. In quantum systems, the quantum fluctuations, which prevail at sufficiently low temperature, can be responsible of interesting phase transitions at T = 0 by varying some parameters. These transitions can be compared with transitions in temperature of classical systems and, in some cases, a rigorous equivalence exists between the T = 0 critical behavior of a d-dimensional quantum system and the $T \neq 0$ critical behavior of corresponding d + 1 classical system [1]. This can be understood from the fact that a quantum hamiltonian contains its own dynamics (due to the non-commutativity of operators) and thus the time plays the role of an extra dimensionality. However in some other cases such kind of equivalence does not hold or is not known and the ground state of the quantum hamiltonian is specific of its quantum nature. The exact solutions for quantum spin systems exist only for very specific cases in one dimension [2]. Therefore there is a need for calculational methods especially suited for quantum systems and not limited to d = 1.

We present here a real-space renormalization-group method able to construct approximately the low lying states of a quantum system and to derive any mean value of interest in the ground state such as the order parameter or the correlation functions. This method was first developed by field theorists who studied several quantum field theory problems on lattices [3]. This method is derived from the pioneering work of WILSON [4] on solving the Kondo problem. We have extended this method for a large variety of uniform quantum spin systems like a one dimensional analog of the Kondo-lattice problem [5], the Ising model with a transverse field in one dimension [6], on bipartite lattices in two dimensions [7] on the triangular lattice with antiferromagnetic couplings [8] and on ramified comb-like structures [9], and the isotropic XY model with a transverse field in one dimension [10] and on several lattices in two dimensions [11]. Applications to disordered systems [12] and to Yang-Lee edge singularity problem [13] were made. Other applications to Fermion systems are expected, already the exact Kondo lattice hamiltonian has been studied in one dimension [14] and the comparison with the Hubbard chain will be made [15]

We have chosen here to present the method when applied to a specific two dimensional problem which considers the extension of the concept of frustration to the case of quantum systems [16]. A generalized spin 1/2 XY model can be written as :

$$H = -\frac{J}{2}\sum_{<ij>}\{\cos\theta(S_i^x S_j^x + S_i^y S_j^y)+\sin\theta(S_i^x S_j^y - S_i^y S_j^x)\} - h\sum_i S_i^z$$

*Supported by the Deutsche Forschungsgemeinschaft Bonn.

where $h > 0$, $J > 0$ and θ is an angular parameter. The nearest neighbor interaction considered here is a quantum extension of the classical interaction $J\cos(\theta-\phi_{ij})$ which is maximum when the angle ϕ_{ij} between neighboring spins is θ.

For $h = 0$, this model includes ferro- and antiferromagnetic XY models ($\theta = 0$ and π respectively) as well as Dzialoshinksy-Moriya model ($\theta = \frac{\pi}{2}$). The study of (1) is particularly interesting on a non-bipartite lattice, such as the triangular lattice, which gives rise to frustration effects for classical antiferromagnetic interactions. Here the frustration can be introduced gradually by varying θ. For $\theta \neq 0$ no exact informations about the ground state are available. Only very recently some indications for frustration effects were obtained for $\theta = \pi$: by extrapolating from finite cell calculations[17] an important increase of the ground state energy compared to the ferromagnetic case was observed. Another indication is the exact value for the critical field above which the gap opens [16] :

$$(h/J)_c = 3\cos(\theta - 2n\pi/3) \quad \text{for} \quad (2n-1)\pi/3 < \theta < (2n+1)\pi/3 \tag{2}$$

by going from ferro- to antiferromagnetic situation one observes a strong reduction of the critical field $\left(h_c(\pi) = 0.5\, h_c(o)\right)$

Before presenting the renormalization group method applied to (1) on the triangular lattice let us precise that the bonds have been oriented as in fig. 1 to define unambiguously the sign of the antisymmetric part of (1). Moreover all the properties of (1) depend on θ with period $2\pi/3$: by rotating the spins on sublattices A, B, C of fig. 1 by respectively 0, $+ 2\pi/3$, $-2\pi/3$, (1) remains unchanged after changing θ in $\theta + 2\pi/3$. In particular $\theta = \pi/3$ is equivalent to the antiferromagnetic case $\theta = \pi$, so that by simply varying θ from 0 to $\pi/3$, we go continuously from the unfrustrated to the fully frustrated situation.

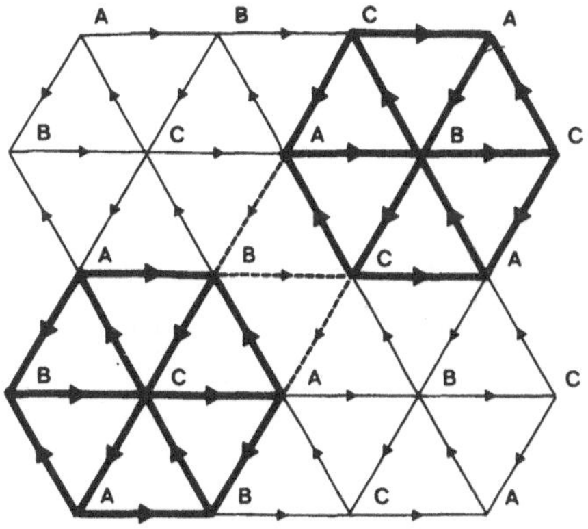

Fig.1 Two adjacent blocks used with their intrablock (thick lines) and interblock (dashed lines) couplings

The renormalization-group method can be considered as an iterative and approximate construction of the ground state of (1) for the infinite two-dimensional triangular lattice. At each iteration the hamiltonian takes the form :

$$H = -\sum_{<i,j>} \{J_1(S_i^+S_j^- + S_i^-S_j^+) + i\, J_2(S_i^+S_j^- - S_i^-S_j^+)\} - h\sum_i S_i^z \tag{3}$$

where

$$S_i^{\pm} = \frac{1}{2}(S_i^x \pm i\, S_i^y) \tag{4}$$

and

$$J_1 = J\cos\theta$$
$$J_2 = J\sin\theta\,. \tag{5}$$

The triangular lattice is split into adjacent blocks of seven sites with one central spin labelled p = 0 surrounded by six peripherical spins labelled p = 1, .., 6.

The centers of the blocks form a new lattice dilated by $\sqrt{7}$ compared to the original one. The first step of the renormalization group procedure is to diagonalize exactly the intrablock part of hamiltonian (3). A number n_L of lowest lying states of a block is retained. They form a new truncated basis in which the new parameters of the system are recalculated by pulling back the initial interblock couplings. The procedure is then iterated until no change in the parameters are produced i. e. until a fixed point is reached. We shall limit ourselves in this study by keeping only two ($n_L \equiv 2$) lowest lying states of a block. Calling them $| + >'$ and $| - >'$ and their energies E_+ and E_- we can introduce a new spin $\underset{\sim}{S}'$ for the block, the eigenstates of S'^z being respectively $| + >'$ and $| - >'$. The search of these states is simplified by considering the block symmetry and by considering that $\sum_{p=0} S_p^z$ commutes with the block hamiltonian so that the eigenstates can be classified in term of the different eigenvalues + 7, + 5 ... - 7 of this operator. By varying h we observe some crossing in the low lying levels so that we can define different possible renormalization group transformations. We have considered two cases : (a) For h = 0, the two lowest states $|+>'$ and $|->'$ form a doublet ($E_+=E_-$) belonging respectively to $\Sigma S_p^z=+1$ and -1. (b) For sufficiently large h value, the two lowest states $|+>'$ and $|->'$ belong respectively to $\Sigma S_p^z = +7$ and +5. In both cases the renormalization group equations take the same qualitative form. By rewriting the block hamiltonian we obtain the new field h' :

$$h' = \frac{1}{2} (E_- - E_+) \tag{6}$$

Doing so, we drop a constant term $\frac{1}{2}(E_+ + E_-)$ which will be useful in calculating the ground state energy per site. To rewrite the interblock coupling, we first express the old spin operators $S_p^{\pm}$ in terms of the new block spin operators $S^{\pm\prime}$. We obtain the following spin recursion relations

$$S_o^+ = \xi_o \, S^{+\prime} \tag{7}$$

$$S_p^+ = (\xi_R + (-1)^p \, i \, \xi_I) \, S^{+\prime} \qquad \text{for } p = 1, \ldots, 6 \tag{8}$$

and complex conjugate equations for S_p^-. In both cases the real parameters ξ_o, ξ_R and ξ_I can be calculated as a function of J_1, J_2 and h. Then, (8) allows us to rewrite the interblock coupling which takes the same form as in (3) but with new constants J_1' and J_2' given by :

$$\begin{aligned} J_1' &= (3\,\xi_R^2 + \xi_I^2\,)\,J_1 + 2\xi_R\,\xi_I\,J_2 \\ J_2' &= 3\xi_R\,\xi_I\,J_1 + (\xi_R^2 + 3\xi_I^2)\,J_2\,. \end{aligned} \tag{9}$$

We can then calculate at each step the new parameters J' and θ' by inverting (5). Eqs. (6) (9) form a set of renormalization-group equations which can be iterated up to a fixed point.

Before discussing the renormalization group trajectories in each case let us show how one can calculate the ground state energy per site.

This quantity is evaluated by cumulating the constant terms $(E_+ + E_-)^{(n)}/2$ divided by the number of sites at each step n, which is 7^{n+1} :

$$- E/N = \frac{1}{2} \sum_{n=0}^{\infty} (E_+^{(n)} + E_-^{(n)})/7^{n+1}\,. \tag{10}$$

In the case of small compact block, as that considered here, which does not respect the ratio b_s between the number of bonds and sites on the infinite lattice, (10) gives only a lower bound estimation. An alternative (upper bound) estimation is obtained by dividing instead by the number of bonds at each step $3.7^{n+1} - 3^{n+2}$, and then multiplying the result by $b_s = 3$

$$- E'/N = \frac{1}{2} \sum_{n=0}^{\infty} (E_+^{(n)} + E_-^{(n)})/(7^{n+1} - 3^{n+1}) \ . \tag{11}$$

We have observed that the arithmetical average between these two estimations $- E/N = - \frac{1}{2}(E + E')/N$ gives good results for several two dimensional lattices when comparing with other estimations [11].

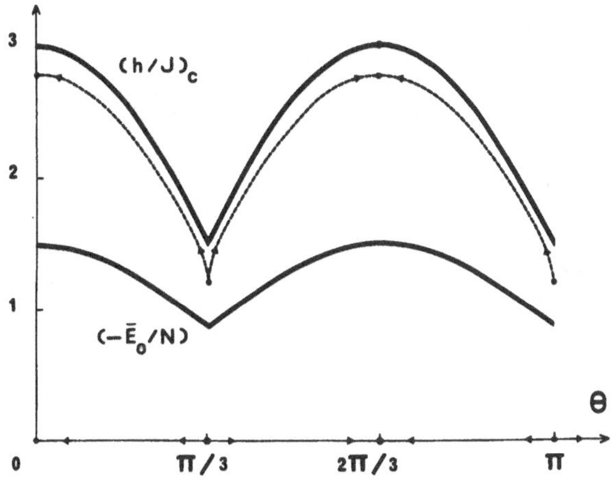

Fig.2 The renormalization group results for the critical field (dashed upper curve) compared with the exact results(full upper curve) and for the ground state energy per site for h=0

Let us now present the results. For any set of initial values $h^{(o)}$, $J^{(o)}$, $\theta^{(o)}$ of h, J, θ we observe that $\theta^{(n)} \to 0 \pmod{2\pi/3}$ except if $\theta^{(o)} = \pi/3 \pmod{2\pi/3}$ where $\theta^{(n)}$ remains constant. Thus, in the case (a) where θ is the only dimensionless parameter, the ferromagnetic ($\theta = 0 \bmod 2\pi/3$) and antiferromagnetic ($\theta = \pi \bmod 2\pi/3$) situations are respectively stable and unstable fixed points. In the case (b) we can determine a critical curve $(h/J)_c$ reported in fig. 2 (dashed curve). Starting with $h^{(o)}/J^{(o)}$ different from $(h/J)_c$ we always flow away from this critical line. The comparison with the exact curve given by (2) (full curve) gives an idea of the approximation of the method.

At each fixed point, we can evaluate the dynamical exponent z which tell us how the energy scales with length. z is calculated directly from the renormalization of J at the fixed point : $z = - \log (J^{(n+1)}/J^{(n)})/\log \sqrt{7}$. In both cases (a) and (b) z is found to be larger for the unstable fixed point (antiferromagnetic situation).

At last we have calculated the ground state energy per site in case (a) (full lower curve). We find $- \bar{E}/N \simeq 1.5$ and 0.88 for $\theta = 0$ and $\theta = \pi$ respectively in reasonable agreement with previous estimations [17].

As a general conclusion we can precise the effect of frustration in this quantum system : there is a general shrink of the low energy spectrum reflected by a strong reduction of the critical field above which a gap opens and an important increase of the ground state energy. Moreover the fully frustrated antiferromagnetic case has scalling properties peculiar to its own nature, while any other situation has scalling properties of the ferromagnetic case. The larger exponent z in this case is characteristic of a large density of states near the bottom of the energy spectrum. These effects are reminiscent of the effects of frustration in classical systems.

References

1 M. Suzuki, Prog. Theor. Phys. 46, 1337 (1971) ; Prog. Theor. Phys. 56,1454 (1976)
2 C. J. Thompson, in "Phase transitions and critical phenomena", vol. I, C. Domb and M. S. Green ed. Academic Press, 1972
3 S. D. Drell, M. Weinstein and S. Yankielovicz, Phys. Rev. D 14 487 (1976) ; 14, 1628 (1976) ; 16, 1769 (1977)
S. D. Drell ; M. Svetitsky and M. Weinstein, Phys. Rev. D 17, 523 (1978)
S. D. Drell and M. Weinstein, Phys. Rev. D 17, 3203 (1978)
4 K. G. Wilson, Rev. Mod. Phys. 47, 773 (1975)
5 R. Jullien, J. N. Fields and S. Doniach, Phys. Rev. Lett. 38, 1500 (1977) ; Phys. Rev. B 16, 4889 (1977)
6 R. Jullien, P. Pfeuty, J. N. Fields and S. Doniach, Phys. Rev. B 18, 3568 (1978)
7 K. A. Penson, R. Jullien and P. Pfeuty, Phys. Rev. B 19, 4653 (1979)
8 K. A. Penson, R. Jullien and P. Pfeuty, J. Phys. C 12, 3967 (1979)
9 R. Jullien, K. A. Penson and P. Pfeuty, J. Physique (France) 40, L 237 (1979)
10 R. Jullien and P. Pfeuty, Phys. Rev. B 19, 4646 (1979)
11 K. A. Penson, R. Jullien and P. Pfeuty, Phys. Rev. B (1980)
12 K. Uzelac, K. A. Penson, R. Jullien and P. Pfeuty, J. Phys. A 12, L 295 (1979)
13 K. Uzelac, P. Pfeuty and R. Jullien, Phys. Rev. Lett. 43, 805 (1979),
K. Uzelac, R. Jullien and P. Pfeuty, Phys. Rev. B (1980)
K. Uzelac, R. Jullien, P. Moussa and P. Pfeuty, this conference
14 R. Jullien, P. Pfeuty, A. K. Bhattacharjee and B. Coqblin, J. of Applied Phys. 50, 7555 (1979)
15 P. Pfeuty, R. Jullien and K. A. Penson, in preparation
16 R. Jullien, K. A. Penson, P. Pfeuty and K. Uzelac, Phys. Rev. Lett. to be published.
17 L. G. Marland and D. D. Betts, Phys. Rev. Lett. 43, 1618 (1979)

Yang-Lee Edge Singularity by Real Space Renormalization Group

K. Uzelac+, R. Jullien, P. Pfeuty

Laboratoire de Physique des Solides, Bâtiment 510, Université Paris Sud
Centre d'Orsay
F-91405 Orsay, France

P. Moussa

Service de Physique Théorique, C.E.A., Orme des Merisiers
F-91190 Gif sur Yvette, France

Quite a long time ago Yang and Lee [1] pointed out the importance of the relationship between the zeros of the partition function and the singularities of thermodynamic quantities occuring in a second order phase transition. They also proved the theorem that for the ferromagnetic Ising model these zeros lie on the unit circle in the complex activity plane $z = \exp(-2h/kT)$ where h is a complex symmetry breaking field and T is the temperature. In the thermodynamic limit they are distributed by some density $g(h)$. Above T_C this distribution has a gap around the real axis, which closes when approaching T_C. Since $g(h)$ is proportional to the spontaneous magnetization, the edges of this gap are branching points for the magnetization.

The work of Yang and Lee was followed by many further analysis of the function $g(h)$ and by generalizations of the Yang and Lee theorem to some other models of phase transitions [2].

Recently another aspect of this problem came out. Namely, the branching point itself can be viewed as a critical point. It is associated with a second order phase transition in the imaginary field with the density of zeros as an order parameter. The concept of universality involving the critical exponents and scaling relations is then introduced. For example, the critical behaviour of $g(h)$ is described by a characteristic exponent σ defined through the relation $g(h) \sim (h - h_c)^{\sigma}$.

This transition is however more complex to study than the "ordinary" one (in zero field), and the exact results are few (1d Ising model [1], spherical model [3], Bethe lattice [4]). In particular, the 2d Ising model which has been solved exactly for an "ordinary" transition, was not solved with an imaginary field. Only the results obtained from high temperature series expansion are available [5][6]. The subject of this talk is to present a real space renormalization group approach to this particular problem. The renormalization group was already applied to the Yang-Lee edge singularity problem by Fisher [7]. He used the Ginzburg-Landau functional formulation of the problem to show that the upper critical dimensionality was 6 and made a $6-\varepsilon$ expansion.

Our renormalization group method is not applied directly to the classical 2d Ising model, but to his onedimensional quantum equivalent. The quantum model considered here is defined by the Hamiltonian

$$H = - \sum_i (J\, S_i^x\, S_{i+1}^x + \Gamma\, S_i^z + ih\, S_i^x), \qquad (1)$$

where S^x, S^z are Pauli matrices. Due to quantum fluctuations, this model exhibits a phase transition at $T = 0$, the ground state changing from doublet to singlet. For $h = 0$ this model is exactly soluble [8] and it has been shown [9] that his critical behaviour is equivalent to that of the classical 2d Ising model. The transverse field Γ and the ground state correspond to the temperature and the free energy of this later respectively. The additional symmetry breaking field ih has the same role in the two models.

+Permanent address : Institute of Physics of the University, Zagreb, Croatia, Yugoslavia

We have applied to a Hamiltonian (1) a T = 0 real space renormalization group method which has been successfully applied to several quantum systems [10].

We start by dividing our 1d system into blocks of n_s spins. In the Hamiltonian (1) one can then distinguish intrablock and interblock parts, i. e.

$$H = \sum_j H_j + H_{j,\,j+1} \quad ,$$

where

$$H_j = - J \sum_{p=1}^{n_s-1} S^x_{j,p} S^x_{j,p+1} - \sum_{p=1}^{n_s} (\Gamma S^z_{j,p} + ih S^x_{j,p}),$$

$$H_{j,\,j+1} = - J S^x_{j,n_s} S^x_{j+1,\,1} \quad ,$$

and j labels the blocks.

The intrablock part is solved exactly. This gives 2^{n_s} eigenstates and corresponding eigenvectors. In order to reproduce the Hamiltonian of the form (1) we make an approximation retaining only the two lowest energy levels. This is justified by the fact that, since we are looking for the ground state, the higher energy levels are assumed not to be strongly relevant for the problem. Then we can define the new operators $\tilde{S}^z_j$, $\tilde{S}^x_j$ in the new basis, and H_j becomes a local term in this new block spin variables

$$H_j = - \tfrac{1}{2}(E_- - E_+) \tilde{S}^z_j + \tfrac{1}{2}(E_+ + E_-) I_j \ . \tag{2}$$

It remains to represent the intrablock interaction H_{ij} in the new basis by expressing the old spins in terms of the new ones

$$S^x_{j,\,n_s} = A \tilde{S}^x_j + B \tilde{S}^z_j + C I_j \ . \tag{3}$$

After a suitable notation, the block-spin Hamiltonian is reduced to the form (1) with new parameters $\tilde{J}$, $\tilde{\Gamma}$, $\tilde{h}$ and an additional constant term $C \sum_i I_i$ which redefines the ground state energy.

Before proceding to the study of the recurrence relations, let us make a few comments more. The first is about the approximation. The only approximation is the truncation of several high energy levels. It can be improved in two ways : by increasing the size of the blocks n_s or by taking larger number of levels. In the later case, instead of the Hamiltonian (1), one has to deal with a Hamiltonian in a more general form involving a larger number of parameters

$$H = \sum_i (D_i - A_i B_{i+1}) \quad ,$$

where A_i, B_i, D_i are the matrices of order n_L. The two approaches have been taken, and shall be discussed later with the results.

The second remark concerns the Hamiltonian which is non-hermitian but symmetric. Thus the eigenvectors are orthogonal in the sense of the real product

$$\sum_k \lambda^i_k \lambda^j_k = \sum_k (\lambda^i_k) \ \cdot \delta_{ij} \quad ,$$

and we choose the corresponding norm which is generaly complex. There is also a problem of choice for the lowest energy levels, since they are complex also. In order to obtain an analytic continuation from the case h = 0, we consider as "lowest" those with the lowest real parts.

Let us now turn back to the recurrence relations for $n_L = 2$. The corresponding parameter space is presented in fig. (1). The fixed point of the ordinary transition appears to be unstable with respect to the parameter h/Γ. There is a line

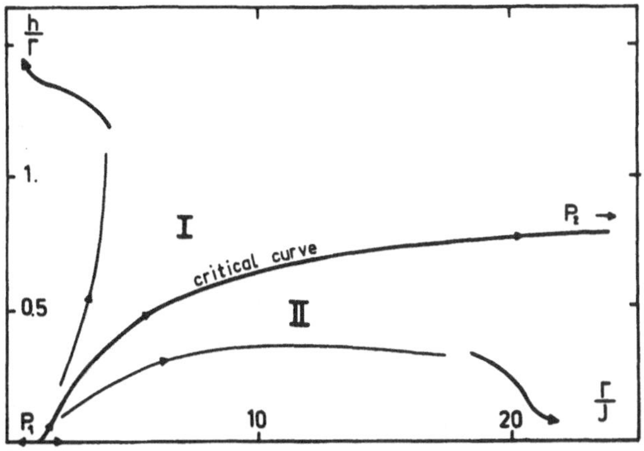

Fig.1 Trajectories in the parameter space for $n_L = 2$, $n_S = 3$. The critical curve separates the ordered phase (I) from the desordered phase (II)

separating the two phases which ends up with a fixed point at $\Gamma/J = \infty$, $h/\Gamma = 1$. This new, once unstable, fixed point governs the imaginary field transition.

The density of zeros is obtained by calculating the x component of the magnetization, since

$$g(h) \sim \mathrm{Re} < S^x > \quad .$$

The magnetization components in the ground state can be evaluated by expressing the original averages $< S_i^x >^{(o)}$ and $< S_i^z >^{(o)}$ as a function of $< S_j^x >^{(n)}$ and $< S_j^z >^{(n)}$ in the fixed point, which are known quantities. The result depends on the set of n values relating i to j, where n is number of steps needed to reach a fixed point. As shown in ref. [10] the best result is obtained when renormalizing a spin in the middle of the block to avoid edge effects.

Fig. (2) represents the real part of $< S^x >$ as a function of h, for three different values of Γ/J : 1.7 ; 1.3 ; $(\Gamma/J)_c + 10^{-3}$, i. e. when crossing the critical line of fig. (1) in three different points. One can observe very pronounced crossover for the point nearest to the "ordinary" transition fixed point.

The exponent σ has been calculated for $n_L = 2$, $n_S = 2, 3, 4, 5$ and for $n_L = 4$, $n_S = 3$. The results for $n_L = 2$ are presented on fig. 3 together with the results for β. The squares and triangles represent the values for different sizes of blocks with odd and even n_S values respectively. For reasons of block symmetry there is

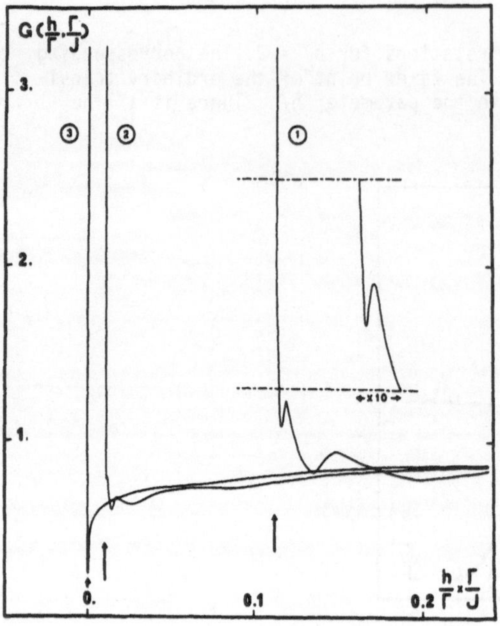

Fig.2 Density of zeros as a function of the imaginary field calculated for $n_L=2$, $n_S=3$. The curves 1, 2, 3 correspond to the $\Gamma/J=1.7$, 1.3 and $(\Gamma/J)_c + 10^{-3}$ respectively

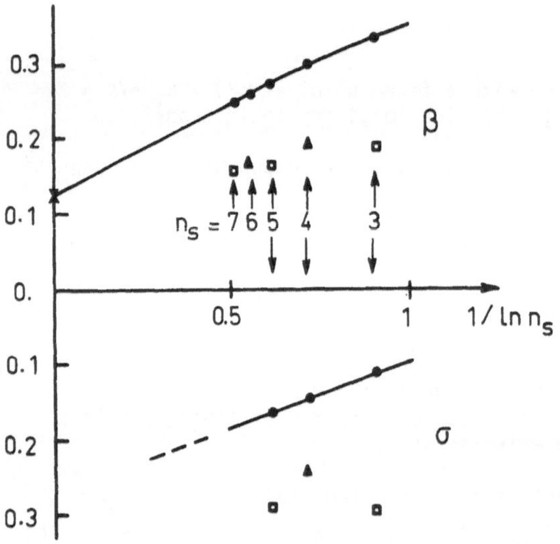

Fig.3 The results for the critical exponent σ and β obtained for $n_L=2$ for different n_S

not a monotonous convergence in function on n_S. In order to follow better the dependence on n_S we have made another calculation of $< S^x >$ where we renormalize the average of S^x over the block. It is represented by dots on fig. (3). In this case the results are obviously worse because of edge effects but they depend monotonically on n_S. The figure shows that the convergence in n_S is very slow (as $1/\ln n_S$). Nevertheless, in the case of β, one can see that the extrapolation fits well with the exact value. Such a, extrapolation for σ will give a value in the interval between - 0.22 and - 0.28, which would lead to a σ value larger in absolute value than the value - 0.16 predicted from series expansion [5][6]. Increasing n_L improves better the results. For $n_L = 4$, $n_S = 3$ we obtain $\beta = 0.124$ ($\beta_{exact} = 0.125$). For σ we obtain - 0.22.

Another effect which can be observed on the fig. (2) are the oscillations in Re < S^x >. They have well defined period, so that the density of zeros can be written as

$$g(\frac{h}{\Gamma}) \sim \left(\frac{h-h_c}{\Gamma}\right)^{\sigma} \times \{A + B\cos(\frac{2\pi}{X} \ln \frac{h-h_c}{\Gamma})\}$$

Such oscillations can occur as an artifact in real space renormalization group procedure, and then their period is $(1/\nu)\ln n_s$. As can be seen from table I this

n_s / $n_L = 2$	2	3	4	5	$n_s = 3$ / $n_L = 4$
$(\frac{h}{J})_c$	1.28	1.15	1.11	1.08	0.97
X	2.2	1.7	1.0	3.1	2.4
$\ell n\lambda_1$	1.08	3.14	3.74	4.53	2.8

Table 1

interpretation of our oscillations was not very convincing for $n_L = 2$, but is acceptable for $n_L = 4$.

In conclusion, the present investigation of the Yang-Lee edge singularity of the two dimensional classical Ising model by a renormalization group calculation done on a one dimensional quantum equivalent appears to be really powerful. It would be interesting to extend the present calculation to higher dimensionalities. For example, it could be possible to estimate σ for the 3d classical Ising model by studying the quantum transverse field Ising model on the honeycomb 2d lattice for which one has already good results without complex longitudinal field [10].

References

1 C. N. Yang and T. D. Lee, Phys. Rev. 87, 404 (1952)
T. D. Lee and C. N. Yang, Phys. Rev. 87, 410 (1952)
2 See for example D. A. Kurtze, thesis and references therein.
3 D. A. Kurtze and M. E. Fisher, J. Stat. Phys. 19, 205 (1978)
4 G. A. Baker, M. E. Fisher and P. Moussa, Phys. Rev. Lett. 42, 615 (1979)
5 J. P. Kortman and R. B. Griffiths, Phys. Rev. Lett. 27, 1439 (1971)
6 D. A. Kurtze and M. E. Fisher, preprint
7 M. E. Fisher, Phys. Rev. Lett. 40, 1610 (1978)
8 P. Pfeuty, Ann. Phys. (N. Y.) 57 (1970)
9 M. Suzuki, Prog. Theor. Phys. 38, 1225 (1967) and 56, 1454 (1976)
10 R. Jullien, P. Pfeuty, J. N. Fields and S. Doniach, Phys. Rev. B 18, 3568 (1978) and see also refs. 3-13 in K. A. Penson, R. Jullien, P. Pfeuty, K. Uzelac, this conference

Chapter 3
Applications in Biology

Numerical Determination of a Periodical Solution of Discontinuous Type, near a Singular Point, for a Neurophysiological Model

J. Argemi
LMA, CNRS, 31 ch. J. Aiguier, F-13274 Marseille Cêdex, France
B. Rossetto
Université de Toulon et du Var, 83130 La Garde, France

The model considered herein concerns slow potential waves (or burst-plateau patterns) produced by Ba-treated *Aplysia* neurons. The neuronal membrane is traversed by a constant current (so-called the external current), perhaps null.

We are therefore brought to study an autonomous dynamical system, defined in R^4 and including a small parameter ε in two velocity components. From the qualitative results obtained in the particular case where the external current is null, we can envisage here, when $\varepsilon \to 0+$, a numerical study of this system. By this means, a stable periodical solution of a discontinuous type can be determined, which can coexist with a singular point, also stable, in absence of all other limited unstable solutions. This particular type of bi-stable functioning is obtained from a bifurcation (current) value corresponding to the presence of a singular point of the dynamical system on the fold of surface S (similar to the cusp surface) on which the slow movement is defined. Finally, the bifurcation values can be determined for which there is successively the appearance of an unstable periodical solution (which arises in crossing a "pseudo-singular" point located on the fold of S) and then fusion of the latter with the stable periodical solution.

References

1. Argemi: Equa. diff. 78 – Convegno Int. su Equazioni Differenziali Ordinarie, Firenze (Italy) pp.333-340
2. Argemi, Gola, Chagneux: Bull. Math. Biol. *41*, 665-686
3. Argemi, Gola, Chagneux: Bull. Math. Biol. *42*, 221-238

On the Relation Between the Logical Structure of Systems and Their Ability to Generate Multiple Steady States or Sustained Oscillations

R. Thomas

Université Libre de Bruxelles, Laboratoire de Génétique
67, Rue des Chevaux, 1640 Rhode Saint Genèse, Belgium

Summary

Simple feedback loops behave in two essentially different ways depending on whether they contain an *odd* number of inhibitory elements ("negative" loops) or an *even* number of inhibitory elements ("positive" loops); for proper values of parameters or delays, the former generate sustained oscillations, the latter, multiple steady states. For more complex systems, as far as one can tell, the presence of at least one negative loop in the logical structure appears as a necessary (but not sufficient) condition for a permanent periodic behaviour, and the presence of at least one positive loop as a necessary (but not sufficient) condition for multiple steady states.

The second part of this paper deals with the cooperative use of Boolean and continuous methods in the field and with the relations between Boolean and quantitative iteration methods.

1. General Aspects

In various disciplines one has to deal with regulatory systems in which elements take part in the control of their own future production. For instance, substance α may influence (catalyze or inhibit) the synthesis of substance β, which influences the synthesis of substance γ, which in turn influences the synthesis of substance α; this is a typical *feedback loop*. Such situations are found, for example, in genetics, where elements α, β and γ are gene products. In the simple case just mentioned, the rate of synthesis of each element is directly controlled only by the concentration of the preceding element in the loop but it is also true that each element takes part in the control of its own future production, indirectly through the other elements of the loop.

Usually, the actual situations are much more intricate, and it is by no means obvious how one can answer the questions: a) Given a logical structure, which dynamic behaviour(s) does it authorize — more specifically, what are the possible final states, by what pathways are they reached, and what conditions determine which pathway will be followed? b) Conversely, given a dynamic behaviour and assumptions regarding the elements responsible for this behaviour, to what extent can one proceed rationally towards the identification of the type of circuit involved? The dynamics of the systems we are dealing with can be analyzed using continuous [6,7,8] or logical [9,10,11] methods (for combined methods see [32-37]). One should not expect complete agreement between the results of continuous and discrete methods, in view of the different simplifying assumptions used. Nevertheless, the agreement is good on the essential points. When a same structure is described in continuous and in discontinuous (logical) terms, the main difference is that the domain of the space of parameters ascribed to the "non-trivial" behaviours tends to be exaggerated by the logical analysis (in which each interaction is treated as if it were infinitely non-linear) and it may be underestimated by the continuous analysis (mainly because absolute time lags, which exist in many systems, are usually not considered [12]). Although the detailed mechanisms are extremely diverse, and the very nature of the elements and interactions involved is different in various disciplines, general ideas have emerged. Of utmost importance is the discovery that such non-trivial behaviours as multiple steady states or sustained oscillations can take place far from equilibrium in the presence of appropriate non-linear interactions [2,3]. In biological systems, stable situations are indeed almost always stationary states rather than equilibria, and non-linearities are often provided by oligomeric proteins, whose monomers act in a cooperative way [4,5].

A significant recent advance is that one begins to perceive the relationship between *logical structures* and dynamical behaviour of systems. Specific situations are dealt with elsewhere. Here, I wish rather to express very general considerations regarding the relationship between the logical structure and the dynamical behaviour of systems comprising feedback loops, with special reference to the question of which logical structures can, and which cannot, generate bifurcations leading to multiple steady states or sustained oscillations.

1.1 Two Types of Regulation

In spite of the variety and complexity of regulatory systems, it is nevertheless possible to distinguish two fundamentally different types of control. In one type, one deals with a *normative* mechanism, that is a mechanism which tends to keep the value of a variable between certain limits by applying a correction each time it departs too much from a fixed value. Familiar examples are the Watt regulator for steam machines, and thermostats — the natural ones which ensure thermal homeostasis in mammals and birds, as well as man-made thermostats. The result of this type of regulation is a *periodic situation*. The amplitude depends on various factors, including the times of response of the elements of the system; under certain conditions, for example if these times are very small, the amplitude of the oscillation may vanish, thus leading in practice to a stable steady state. But even in this case one should keep in mind that in these systems stability is obtained by repeated corrections acting alternately upwards and downwards.

Another type of control endows systems with the ability *to choose between two or more attractors*. Examples of fundamental importance are found in the development of living organisms; cells carrying the same genetic information and subject to the same present environment can be permanently locked in distinct states of differentiation as a result of differences in their former history. This type of control deserves a special interest because, in contrast with the first type examplified by thermostats, on the one hand it can generate diversity, and on the other hand once the choices have been made the system tends to remain stable; occasional perturbations can be corrected if they are not too ample but the stability of the system does not depend upon the repeatedly alternating corrections described above.

1.2 Simple Feedback Loops

One can denote by *simple feedback loops* closed logical structures in which each element is under the direct control only of the element upstream and directly controls only the element downstream (as in the example in the first paragraph of this paper). For proper values of the parameters, these simple logical structures can generate either of two non-trivial behaviours: sustained oscillations or multiple steady states. Which of these behaviours takes place depends only on the *parity* of the number of *negative* (inhibitory) elements in the loop (see, for instance, [11,13]). If the loop comprises an *odd* number of elements, each element exerts (via the other elements in the loop) a negative effect on its own production, and for proper values of the parameters this results in a sustained periodic behaviour. Let us consider, for instance, three interacting genes A, B and C, and suppose that α, the product of gene A, is necessary for the expression of gene B, whose product β prevents the expression of gene C, whose product γ in turn is necessary for the expression of gene A. When gene A has been "on" for a while, the concentration of its product α becomes sufficient to switch on gene B, whose product β will switch off gene C; and as the product γ of gene C has decayed (or has been diluted out) gene A will be switched off, etc...

The loops comprising an *even* number of negative elements function in a radically different way. It can be shown that each element exerts (via the other elements in the loop) a positive control on its future production. In other words, an element tends to be present or absent to the extent that it was already present or absent before. The simplest case of this type of control is pure autocatalysis (a loop with only one, positive, element); if the production of a substance requires that this substance be present above a threshold concentration, either the substance was already present and in proper conditions it will continue to be synthesized and remain present, or it was not present and in otherwise identical conditions it will not be synthesized and will remain absent indefinitely. As pointed out by DELBROCK [14] more than 30 years ago, multiple steady states can also be produced by two mutually inhibitory elements, either of which will be active in a stable way. More generally, simple feedback loops comprising an even number of negative elements permit a *choice* between two stable steady states (and an unstable situation whose maintenance would require non-realistic conditions: see [1], 125-6).

As the control exerted by each element on its own future production is negative when there is an odd number of inhibitory elements in the loop and positive when the number is even, it is appropriate to denote the two types of loops "negative loops" and "positive loops", respectively. Using this nomenclature, one can thus briefly restate the content of the preceding paragraphs by saying that *for appropriate values of the parameters or delays, negative loops will generate a permanent periodic behaviour*, while *positive loops will permit a choice between two stable states*. Note that with only negative elements one can build positive as well as negative loops, while with only positive elements one can build only positive loops.

1.3 Systems with Interconnected Loops

Simple feedback loops are ideal objects. In practice, one has usually to consider nets comprising several interconnected loops. Within such a net, one can formally distinguish individual loops by following the oriented graph in an appropriate way, and each of these loops is positive or negative according to the criterion of the parity of the negative elements. However, these individual loops are usually no longer simple loops because their elements may have more than one input. As a result, the behaviour which would characterize the corresponding simple loop may be lost. Consider, for instance, a simple negative loop in which element α is necessary for the synthesis of β, element β inhibits the synthesis of γ and γ is necessary for the synthesis of α. For proper values of the parameters such a system will undergo sustained oscillations; however, grafting an additional control (the synthesis of α requires γ *and* α *itself*) abolishes this possibility, and the only final state is now a stable state in which the syntheses of α and β are off and the synthesis of γ is on.

Various individual combinations of positive and negative loops are analyzed elsewhere [11,25] but can one tell something more general? As far as one can extrapolate from the hundreds of networks analyzed so far, *the presence of a negative loop in the logical structure of a system is a necessary, although not sufficient, condition for a permanent periodic behaviour, and the presence of a positive loop is a necessary, although not sufficient, condition for multiple stable steady states.*

1.4 Real Systems

In this perspective, it is instructive to examine in various fields what kinds of models have been proposed to account for experimentally observed cases of "non-trivial" behaviour. Multiple steady states ("epigenetic" differences) have been experimentally observed in bacteria [15,16,17,18], and have been shown to be due to autocatalysis or to mutually inhibitory gene products. Sustained oscillations have been studied experimentally and theoretically, in particular in the case of glycolysis [19,20] and the

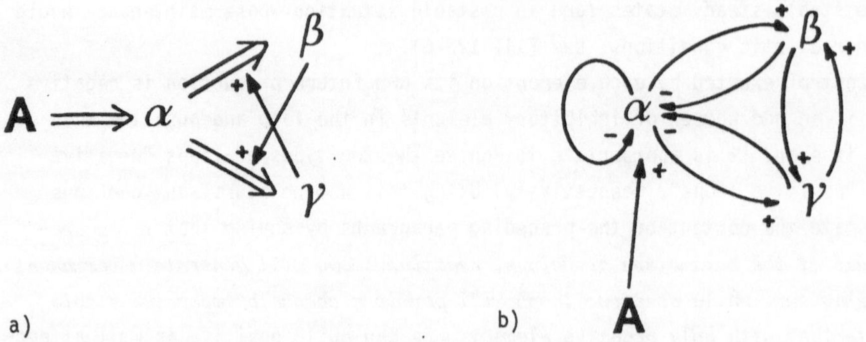

a) b)

Fig. 1. (a) A *scheme* in which the transformation of substances into one another is symbolized ⇒ and the catalytic effect of a substance on a reaction is symbolized →+. (b) The corresponding *logical structure*, in which one sees not only the positive actions of α towards β and γ, and of β and γ towards each other, but also the negative actions of α, β and γ on α. The nature (whether "and" or "or") of the connections has not been indicated in the graph but it is entirely explicit in the set of logical equations:

$$a = A(\bar{\alpha}+\bar{\beta}\cdot\bar{\gamma}) \quad , \quad b = \alpha\cdot\gamma \quad , \quad c = \alpha\cdot\beta$$

rhythmic synthesis of cAMP in *Dictyostelium* [21,22]. As regards the models put forward to account for stable oscillations, probably the most extensively studied one is a particular type of negative loop comprising n-1 positive elements and one negative element [6,7,8]. On the other hand, oscillations can also be accounted for by models whose most apparent feature is a positive loop. For instance, a model proposed by GOLDBETER [22] can be described in a slightly simplified way as follows. A substance α generated by a source A can transform into substances β or γ; the conversion of α into β requires the presence of γ and the conversion into γ requires the presence of β (see Fig.1a).

At first view the model comprises only positive interactions and consequently no negative loop, and this would disagree with our conjecture that a negative loop is a necessary condition for permanent oscillations. However, the apparent discrepancy disappears if one notices that the role exerted by α in the production of β and γ is not a catalytic one but rather a conversion of α into β or γ; the positive actions of α on β and γ, of β on γ and of γ on β, all take place *at the expense of* α. More specifically, the higher the concentrations of α, β and γ, the higher the rates of conversion of α into the other substances; thus, α, β and γ exert a negative effect on the *net* synthesis of α (the resultant of the conversion of A into α and of the conversion of α into the other substances). This can be seen in the differential equations representing the system: α, β and γ each appear in negative terms in the expression of dα/dt. If one takes into account not only the obvious positive effects in Fig.1a, but also the "hidden" negative effects just described, one gets the complete logical structure of the model (Fig.1b). As shown by GOLDBETER [22], such a system will oscillate permanently for proper values of the parameters. It can be shown on the other hand that this behaviour is absolutely dependent on the presence of the (hidden) negative loops in the logical structure of the system. If one suppresses them, for instance, by increasing the flux of $A \rightarrow \alpha$ to such a level that the concentration of α is no more affected by its conversion into β or γ, the oscillation disappears.

In a different domain, neurobiology, periodic situations are ascribed, depending on the case, to the autonomous periodic activity of individual neurones or to groups of neurones connected in such a way that their very interaction generates a periodic activity [23]. As suggested by SZEKELY [24] and supported by the experimental and theoretical work of FRIESEN and STENT [23], one simple mechanism which generates permanent oscillations is recurrent inhibition, in which each neurone in the circuit is inhibited by the neurone upstream in the circuit. Provided the number of elements is odd[1], the system has to oscillate. I have mentioned above the well-studied type of negative loop comprising only *one* negative element; here, one deals with another particular case of negative loop in which *all* the elements are negative and their number is odd. It is interesting, on the other hand, to notice that much effort had been wasted in the hope of accounting for a permanent periodic behaviour with logical structures with *only* positive loops (e.g., reciprocal inhibition: see the discussion in [23,24,25]).

In conclusion, as far as one can tell from the logical analysis of numerous systems, the presence of at least one negative feedback loop is a necessary condition for a stable periodic behaviour and the presence of at least one positive feedback is a necessary condition for multiple steady states. This should be taken into account when one builds models to account for these "non-trivial" situations.

2. Specific Problems and Methods

2.1 From the Logical Structure to the Behaviours

Given the logical structure of a system, what are the possible final states, by which pathways are they reached, and what conditions determine which pathway will actually be followed?

One approach to this type of problems is a "logical" one, which gives a qualitative view of the possible evolutions of the systems (see [1,10,11,26]). The method I developed is partly derived from the work of conceptors of logical machines [27].

1 A circuit comprising an even number of negative elements connected in the "recurrent inhibition" mode would be a positive loop and it would not yield sustained oscillations. In order to permit sustained oscillations, one has to add appropriate cross connections [23,24] which in fact introduce negative loop(s) into the structure. For a systematic method which tells *which* additional connections should be added (or deleted) see [25].

Essentially, it describes systems by sets of logical equations which relate the state ("on" or "off") of the processes to the presence or absence of their products. For instance, the set of logical equations:

$$a = \bar{\beta} \cdot \bar{\gamma}$$
$$b = \bar{\gamma} \cdot \bar{\alpha}$$
$$c = \bar{\alpha} \cdot \bar{\beta} \qquad (1)$$

tells that process a is "on" if, and only if, β and γ (the products of processes b and c, respectively) are both "absent", that is, their concentration is below its threshold of efficiency. Time would seem at first view to be absent from this description. In fact, function a and variable α (and similarly, b and β, etc.) are related in time as follows. When process a has been off (a=0) for a sufficient time, its product α is absent (α=0); if a signal (in the present case, the fact that $\bar{\beta} \cdot \bar{\gamma}$ switches from 0 to 1) initiates process a, α (the product of process a) will appear, but only after a characteristic delay t_α, and it will remain present as long as process a is on; but if a signal (in the present case, the drop of $\bar{\beta} \cdot \bar{\gamma}$ from 1 to 0) switches off process a, the product α will disappear after a characteristic delay $t_{\bar{\alpha}}$. Thus, function a and variable α are temporally related to each other in the same way as a function and its memorization variable in asynchronous sequential automata (cf. [27]). The difference is that here function a and its memorization variable are qualitatively different objects (one represents a *flux*, the other, an *amount*) whereas in classical sequential logic a function and its memorization variable are one and the same thing, seen at different times. This point has two important implications:

a) In this perspective, a function and its memorization variable correspond to two physical concepts whose relation in time is concretely defined. It ensues that the values of the time delays are no more arbitrary; the establishment delay (t_α) is the time between the onset of a process and the moment its product reaches a significant concentration; the decay delay ($t_{\bar{\alpha}}$) is the time between the moment the process has been switched off and the moment the product has dropped below its efficient concentration (as a result of degradation, dilution, diffusion, etc.).

b) As discussed elsewhere [28] and briefly below (Sect.2.3), this permits drawing a useful parallel between logical equations and the kind of differential equations used in chemical kinetics. In fact, time is (implicitly) involved in similar ways in the two types of equations.

Now, the exact rule used to related each of our functions with its memorization variable must be expressed more explicitly. Let us start with a situation in which the function and its memorization variable have the same logical value, 1 or 0 (this means that either the process is "on" and its product present, or the process is "off" and its product absent). If a signal changes the value of the function, the value of the variable will follow after a characteristic time delay; however, if a

$a = 1$
$b = \bar{\alpha}$
$c = \beta + \gamma$

α	β	γ	a	b	c
$\bar{0}$	$\bar{0}$	0	1	1	0
$\bar{0}$	$\bar{0}$	1	1	1	1
$\bar{0}$	1	1	1	1	1
$\bar{0}$	1	$\bar{0}$	1	1	1
1	$\bar{1}$	$\bar{0}$	1	0	1
1	$\bar{1}$	1	1	0	1
1	0	1	(1	0	1)
1	0	0	(1	0	0)

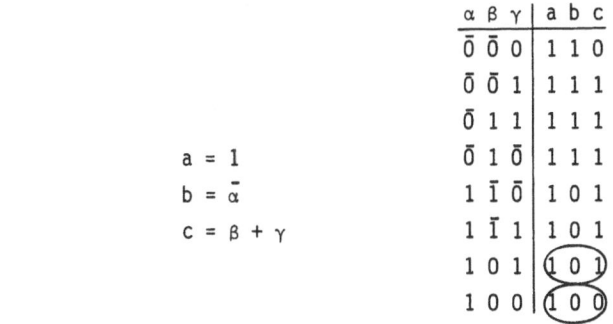

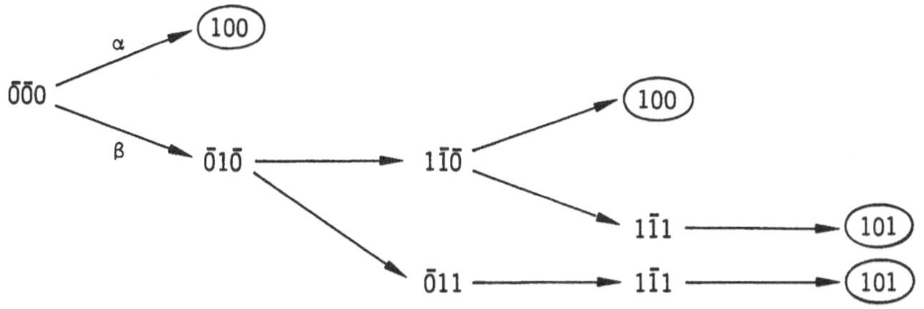

Fig. 2. A simple logical circuit (inspired from the regulation of bacteriophage λ), used to show for a same system the set of logical equations, the states table and the graph giving the temporal sequences of states. Stable states are circled. The condensed formalism ($\bar{0}\bar{0}0$ for 000/110) is used in the graph. The formalism ($\bar{0}\bar{0}0$/110) used in the table is redundant but convenient

counter-order is given before the delay has elapsed, we reason as if the order had never been given. When this rule is used, one deals with an "inertial delay"[2].

Once a model has been translated into a set of logical equations, one can derive a state table which gives the values of the functions for each state (combination of values) of the variables; and from this table one can derive the oriented graph of the Boolean trajectories of the system. It is convenient to treat each state of the variables (α,β,γ,...) and each state of the corresponding functions (a,b,c,...) as "Boolean vectors" (for instance, [34,29]). A state for which the vectors (α,β,γ,...) and (a,b,c,...) are equal is recorded as a *stable state*, as in classical sequential logics. Otherwise, one deals with a transient state; for instance (see Fig.2), if (α,β,γ)/(a,b,c) = 000/110, substances α and β are still absent, but since processes a and b are on, α and β are being synthesized. This can be symbolized in a more condensed way by $\bar{0}\bar{0}0$. The next state will be 100 or 010 depending on whether α or β has reached first its threshold efficiency. In the example chosen, this depends on the

2 This rule has the advantage of simplicity; however for some systems it is somewhat unrealistic because it does not account for the accumulation of a product which would be synthesized in an intermittent way. If one wants to avoid this drawback, other rules can be used (see the work of VAN HAM on "n-order" delays [30]).

relative values of the time delays t_α and t_β. If $t_\alpha < t_\beta$ [3], the system proceeds to the stable state 100; the order "α=1" has been executed and the other "β=1" is cancelled. However, if $t_\beta < t_\alpha$, the system proceeds to the transient state $\bar{0}1\bar{0}$, in which one sees that the order to produce α has not been cancelled but a new order (to produce γ) has been given when β has switched from 0 to 1. Now, the system will proceed to state 110 or to state 011, depending on whether or not $t_\alpha < t_\beta + t_\gamma$; and since these states are themselves transient states ($1\bar{1}\bar{0}$;$\bar{0}11$) the evolution continues. The present system finally proceeds to either of the stable states 100 or 101 depending on the initial state and the values of the time delays.

With this type of analysis, one can treat "by hand" systems with four or a few more variables. For more complex systems programs are available [31]. Note that if nothing is told about the time delays all the pathways of the graph must be considered; if a defined value is ascribed to each time delay, a defined pathway is followed. But when one deals with a population of similar systems, as, for example, a population of living cells, one is usually lead to ascribe to each delay a mean value and a distribution rather than a fixed value. There exist now programs which compute the relative frequencies of the different pathways from the logical equations and the mean value and distribution ascribed to each time delay [31].

2.2 From the Behaviour Towards Logically Acceptable Structures

Inversely, given a behaviour and assuming that the relevant functional elements of the system have been correctly identified, find in a rational way interactions between these elements, which would account for the behaviour. Needless to say, the solutions are multiple; it may nevertheless be useful to identify some of the simplest ones.

I would like to show by an example that this can be done first by a logical method, and if one wants to switch to a quantitative description one can derive from the set of logical equations a homologous set of differential equations which, for proper values of the parameters will provide a qualitatively identical behaviour.

Suppose we want a three-component system which offers a choice between three (and only three) stable states, more specifically those states for which one and only one of the variables is "high" (that is, states 100, 010 and 001). As shown in Fig.3, one can delineate the sets of logical equations which fit the requirements, and among these many possibilities, identify the simplest ones. In this case, the simplest set of logical equations is given by $a = \bar{\beta} \cdot \bar{\gamma}$ and its cyclic permutations. Similarly, the simplest three-variable structure with three stable states such that all but one of the variables are high (110,101,011) is given by equations of the form $a = \bar{\beta} + \bar{\gamma}$. In-

[3] If t_α and t_β happened to have exactly the same value, the transition would be from $\bar{0}\bar{0}0$ to $1\bar{1}\bar{0}$. This possibility is not explicitly mentioned in the graph of Fig.2 but, should it happen, it would automatically be taken into account in the simulations.

(a) General solution

$\alpha\ \beta\ \gamma$	a b c
0 0 0	- - -
0 0 1	(0 0 1)
0 1 1	- - -
0 1 0	(0 1 0)
1 1 0	- - -
1 1 1	- - -
1 0 1	- - -
1 0 0	(1 0 0)

(b) A set of solutions which comprises the symmetrical ones

$\alpha\ \beta\ \gamma$	a b c
$\bar{0}\ \bar{0}\ \bar{0}$	1 1 1
0 0 1	(0 0 1)
0 1 1	- - -
0 1 0	(0 1 0)
1 1 0	- - -
$\bar{1}\ \bar{1}\ \bar{1}$	0 0 0
1 0 1	- - -
1 0 0	(1 0 0)

(c)

a	00	01	11	10	α,β
0	[1]	0	-	[1]	
1	0	-	0	-	
γ					

b	00	01	11	10	α,β
0	[1	1]	-	0	
1	0	-	0	-	
γ					

c	00	01	11	10	α,β
0	[1]	0	-	0	
1	[1]	-	0	-	
γ					

(d) The simplest of the symmetrical solution with no more than 3 stable states

$\alpha\ \beta\ \gamma$	a b c
$\bar{0}\ \bar{0}\ \bar{0}$	1 1 1
0 0 1	(0 0 1)
0 $\bar{1}$ $\bar{1}$	0 0 0
0 1 0	(0 1 0)
$\bar{1}$ $\bar{1}$ 0	0 0 0
$\bar{1}\ \bar{1}\ \bar{1}$	0 0 0
$\bar{1}$ 0 $\bar{1}$	0 0 0
1 0 0	(1 0 0)

(e) Temporal sequences of states in case (d)

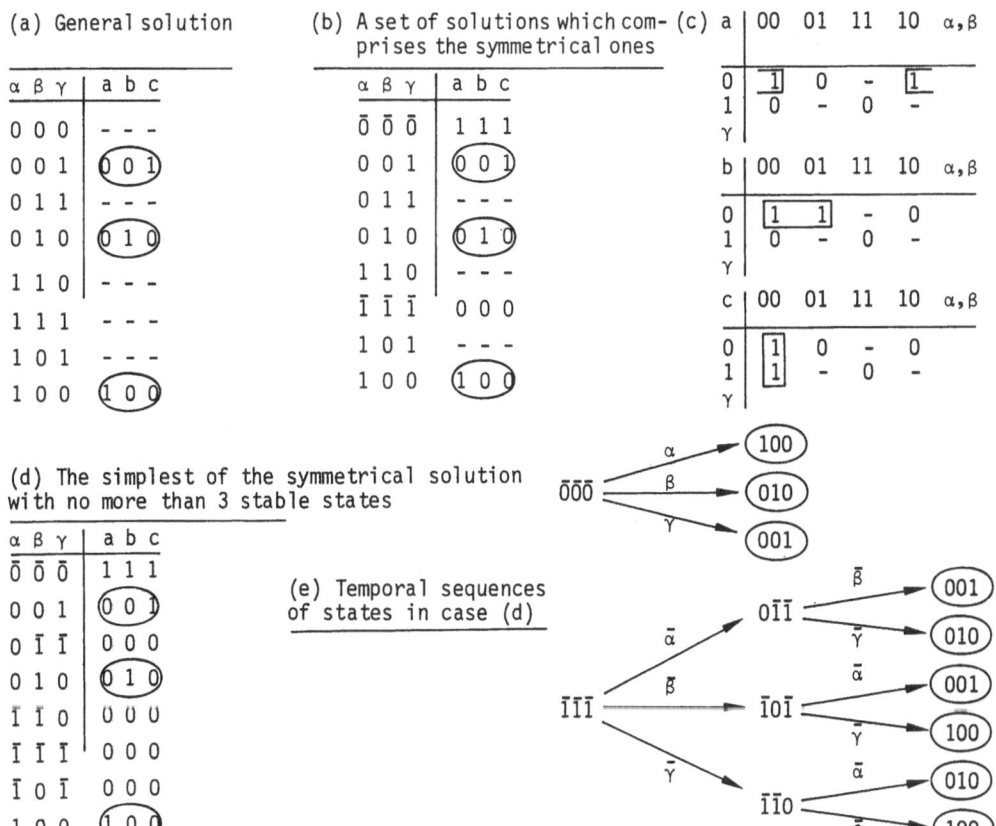

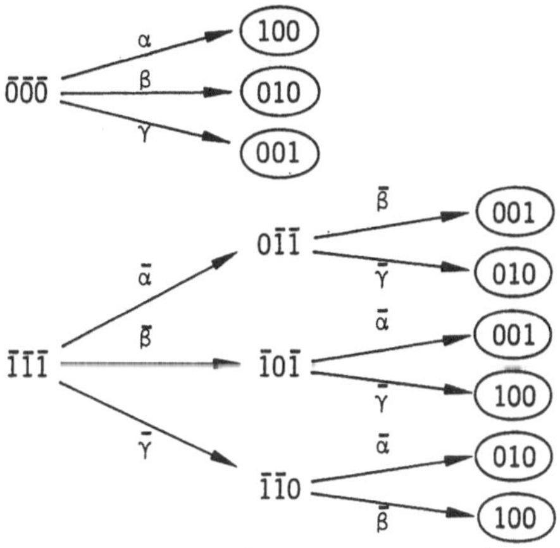

Fig. 3. (Also given as Table 1 in [28].) Derivation of a logical structure permitting a choice between three stable states. (a) This incomplete Karnaugh table mentions the requirements: 001, 010 and 100 should be stable states. Replacing each dot by a 1 or a 0 would provide a complete table consistent with the requirements, and consequently a set of logical equations consistent with the requirements. There are 15 dots, thus 2^{15} ways to fill the table! (b) Among these many possibilities, let us limit ourselves to the subset (b), which comprises all the symmetrical solutions (as well as others). Note that subset (b) still comprises 2^9 solutions. (c) These three tables are just another version of table (b), in which the three functions are presented separately. Under this presentation, one can easily find the prime implicants (see [27,39]) and from these prime implicants the simple solution(s) which fit the desiderata. In this case the simplest set of logical equations is:

$$a = \bar{\beta} \cdot \bar{\gamma} \qquad b = \bar{\gamma} \cdot \bar{\alpha} \qquad c = \bar{\alpha} \cdot \bar{\beta} \ .$$

(d) The Karnaugh table corresponding to the set of equations mentioned under (c). (e) The temporal sequences of states in this same system

cidentally, this can be extended by analogy to systems with n variables: for instance, a system of four logical equations of the form: $a = \bar{\beta} + \bar{\gamma} + \bar{\delta}$ provides a choice between four stable states of the type 0111.

From these sets of logical equations, one can derive sets of differential equations in which the all-or-none interactions are replaced by sigmoids [26]. For instance, one continuous equivalent of (1) is

$$\dot{x} = \frac{1}{1+\alpha_1 y^n} \cdot \frac{1}{1+\alpha_2 z^n} - k_1 x \quad , \text{ etc...} \tag{2}$$

in which x, y, z,... are the continuous variables corresponding to the Boolean variables α, β, γ,... . It can be checked [28] that for proper values of the parameters this system will indeed provide a choice between three stable steady states, depending on the initial state.

Alternatively, once the appropriate logical structure has been found as above, one can use the most elegant method of "piecewise linear equations" developed by GLASS [32,33,34]; this method combines the simplicity of the logical approach with the quantitative character of differential equations.

2.3 On the Relation Between the Logical Equations and the Type of Differential Equations Used in Chemical Kinetics

In equation (2) and, more generally, equations of the form:

$$\dot{x} = f_1(x,y,z,\ldots) - k_{-1}x \tag{3}$$

the first term of the right member expresses the rate of synthesis of the product in terms of the *specific regulatory interactions* of the system. The second term describes the *aspecific decay* of x, which would take place independently of any regulation of its synthesis; this process is usually treated as a monomolecular one ($-k_{-1}x$), a simple and reasonable assumption. Equation (3) simply tells that $\dot{x}$, the *net* rate of variation of x, is the resultant of these two processes. If one rewrites (3)

$$X = \dot{x} + k_{-1}x = f\ (x,y,z,\ldots) \tag{4}$$

one relates the *gross* rate of synthesis of x to the specific regulatory interactions of the system. This is in fact exactly what we do when we write logical equations of the type: $a = \varphi(\alpha,\beta,\gamma,\ldots)$.

Thus, our Boolean function a represents a flux of x, but it is the Boolean equivalent of $\dot{x}+k_{-1}x$ (the gross rate of synthesis), rather than of the time derivative $\dot{x}$ (the net rate of synthesis). In fact, people like geneticists are primarily interested in the gross rate of synthesis of gene products, which parallels the rate of activity of the genes. Besides, the global process usually varies with time in a simpler way than $\dot{x}$ itself. For instance, if a gene, first off, functions at a rate a for a certain period, then is switched off again, $\dot{x}$, initially 0, jumps to a, then decreases progressively to 0 as x (and thus $k_{-1}x$) increases; it jumps to $-k_{-1}x$ when the gene is switched off and progressively raises to 0 as x is degraded; during the same period, X ($=\dot{x}+k_{-1}x$) simply takes the values 0, then a, then 0 again.

Admittedly, in order to follow X as a function of time one has first to calculate $\dot{x}$; but, in the logical method, a, b, c,..., the logical equivalents of X, Y, Z,... can be computed directly as functions of time, in view of the simple rules adopted for the time relation between the functions and their memorization variables. The values of the "aspecific decay constants" k_1 etc. are important as when the gene is switched on they influence the rate of establishment of x and the stationary value of its concentration, and when the gene is switched off, the rate of disappearance of x; in the logical description, they are taken into account in the choice of the values of the establishment and decay delays of each function.

2.4 Iteration Methods

Whether one uses a Boolean or a continuous description, the final states are usually found by some kind of iterative method. For instance, in equations like (3), one reasons that for the steady state(s) $\dot{x}, \dot{y}, \dot{z},\ldots = 0$ and this provides a set of equations of the form $x^0 = 1/k_1\, f_1(x,y,z,\ldots)$ (4).

One can usually reach the stable steady states by very simple iteration methods, as the Jacobi and the Gauss-Seidel methods; it is well known, however, that one must pay attention to possible artefact (absence of convergence in the vicinity of a stable steady state, etc.). In the Jacobi iteration, one introduces initial state values $(x_0, y_0, z_0 \ldots)$ into the right sides of equations (4), calculate the new value of the variables, reintroduce them as x_1, y_1, z_1,... and so on. In the Gauss-Seidel iteration, instead of reintroducing the set of new values as a whole, each new value is reintroduced as soon as it is available (and for this a predetermined order of the variables is used).

Similarly, for our sets of logical equations, one can reason that for the steady states $a = \alpha$, $b = \beta$, etc., and thus $\alpha = \varphi(\alpha,\beta,\gamma)$, and similarly for the other equations. Now, the Boolean values of the steady states can be found by iteration [29].

As noticed by GOLES [38], Boolean iterations can proceed "in parallel" or "in series", and these procedures correspond exactly to the above-mentioned methods of Jacobi and Gauss-Seidel, respectively. In the first case (parallel iteration), from an initial Boolean state $(\alpha_0, \beta_0, \gamma_0 \ldots)$ one computes the values of the functions a, b, c,... which are reintroduced at once as α_1, β_1, $\gamma_1 \ldots$, and so on. In the second case (series), the new value of each variable is reintroduced as soon as it has been calculated, and this is done in a well-defined order, e.g., α, β, γ.

As the reader may have noticed, the way I derive the temporal sequence of the Boolean states is also a kind of iteration. However, the rule is different from either the Jacobi or the Gauss-Seidel iterations. For instance, in the case of Fig.2, from an initial state $\bar{0}\bar{0}0$ the system will proceed to 100 or to 010 depending on whether or not the delay t_α is shorter than t_β; and assuming the second possibility, one will move next from $\bar{0}1\bar{0}$ to 110 or to 011 depending on whether or not $t_\alpha < t_\beta + t_\gamma$, etc.

I have wondered whether one could derive from this third type of logical iteration a more or less analogous numerical method. One simple possibility would consist of calculating from $(x,y,z)_n$ the tentative values $(x,y,z)_{n+1}$ and the differences Δx, Δy, and Δz. Assuming that the fastest process is the one for which Δ is highest, one can use a procedure such that at each round of iteration only the variable for which $|\Delta|$ is highest will switch; but if two or more $|\Delta|$ happen to be identical and higher than the others, the corresponding variables will switch synchronously. The method has been tested on various systems. Its potential interest comes from the fact that in addition to providing (like the other methods) the values of stable steady states, it yields idealized trajectories which are very similar to the Boolean ones. In particular, when there is no stable state it provides a discrete idealization of periodic behaviours which is essentially similar to the Boolean one. It must be clear, however, that the method should be improved, if only by somehow normalizing the Δ.

It may be of some interest to mention that for a given type of iteration one can use programs which, except for the formalization of the functions, are essentially the same, irrespective of whether one uses Boolean or differential equations, or Glass's piecewise linear equations; moderately complex systems can be handled on a pocket calculator.

References

1. R. Thomas (ed.): *Kinetic Logic*, Lecture Notes in Biomathematics, Vol.29 (Springer, Berlin, Heidelberg, New York 1979)
2. P. Glansdorff, I. Prigogine: *Thermodynamics of Structure, Stability and Fluctuation* (Wiley-Interscience, New York 1971)
3. G. Nicolis, I. Prigogine: *Self-Organization in Nonequilibrium Systems* (Wiley-Interscience, New York 1977)
4. J. Monod, J. Wyman, J.P. Changeux: J. Mol. Biol. *12*, 88-118 (1965)
5. M.F. Perutz: Nature *228*, 726-739 (1970)
6. B.C. Goodwin: Adv. Enzyme Regul. *3* (1965)
7. H.G. Othmer: J. Math. Biol. *3*, 53-78 (1976)
8. J.J. Tyson: J. Chem. Phys. *62*, 1010-1015 (1975)
9. S.A. Kauffman: J. Theor. Biol. *22*, 437-467 (1969)
10. R. Thomas: J. Theor. Biol. *42*, 656-683 (1973)
11. R. Thomas: J. Theor. Biol. *73*, 631-635 (1978)
12. J. Richelle: In Ref.[1], pp.281-325
13. P. Van Ham, I. Lasters: J. Theor. Biol. *72*, 269-281 (1978)
14. M. Delbrück: In Unités biologiques douées de continuité génétique (Publications C.N.R.S., 1949) pp.33-35
15. A. Novick, M. Wiener: Proc. Natl. Acad. Sci. USA *43*, 553-566 (1957)
16. M. Cohn, K. Horibata: J. Bacteriol. *78*, 601-612 (1959)
17. H. Eisen, L.H. Pereira da Silva, P. Brachet, J. Jacob: Proc. Natl. Acad. Sci. USA *66*, 855-862 (1970)
18. Z. Neubauer, E. Calef: J. Mol. Biol. *51*, 1-13 (1970)
19. B. Hess, A. Boiteux: Ann. Rev. Biochem. *40*, 237-258 (1971)
20. E.E. Sel'kov: Eur. J. Biochem. *4*, 79-86 (1968)
21. G. Gerisch, B. Hess: Proc. Natl. Acad. Sci. USA *71*, 2118-2122 (1979)
22. A. Goldbeter: Nature *253*, 540-542 (1975)
23. W.D. Friesen, G.S. Stent: Biol. Cybernetics *28*, 27-40 (1977); Ann. Rev. Biophys. Bioeng. *7*, 37-61 (1978)

24. G. Székely: Acta Physiol. Acad. Sci. Hung. *27*, 285-289 (1965)
25. R. Thomas: In Ref. [1], pp.388-399
26. L. Glass, S.A. Kauffman: J. Theor. Biol. *39*, 103-129 (1973)
27. J. Florine: La Synthèse des Machines Logiques (Presses Académiques Européennes, Bruxelles 1964)
28. R. Thomas: Proc. 1980 Solvay Conference in Chemistry, Adv. Chem. Phys. 1982 (in press)
29. F. Robert: Rapport de recherche n°163, IMAG, Grenoble (1979)
30. P. Van Ham: Thesis, University of Brussels (1975)
31. P. Van Ham: In Ref.[1], pp.149-163
32. L. Glass: J. Chem. Phys. *63*, 1325-1335 (1975)
33. L. Glass: In *Statistical Mechanics and Statistical Methods in Theory and Application*, ed. by U. Landman (Plenum, New York 1977)
34. L. Glass, J.S. Pasternak: Bull. Math. Biol. *40*, 27-44 (1978)
35. R.N. Tchuraev, V.A. Ratner: In *Studies on Mathematical Genetics*, ed. by V.A. Ratner (Novosibirsk Inst. Cytol. Genet. Press 1975)
36. V.A. Ratner: *Molekulargenetische Steuerungssysteme* (Akademie, Berlin 1977)
37. A. Rörsch, M.A.E. Groothuis, A.M.H. Schepman: In Ref.[1], pp.440-463
38. E. Goles-Chacc: Rapport de recherches n°157, IMAG, Grenoble (1979)
39. A. Leussler, P. Van Ham: In Ref.[1], pp.62-85

Critical Delays in Logical Asynchronous Models

A. Verhamme and P. Van Ham

Université Libre de Bruxelles, Service des Systêmes Logiques et Numériques, 50, Avenue F.D. Roosevelt, 1050 Bruxelles, Belgium

1. Introduction

The description of complex systems by means of logical variables is a method which allows arriving rapidly at a qualitative understanding of that system [1,2,3]. Such an approach offers a complementary tool to the classical continuous formalism which is more limited in the number of state variables it can handle and is very sensitive to the nonlinear character of the kinetic differential equations. To the reader who is not familiar with the kinetic logic formalization of a model, we give here a short introduction.

The underlying idea is to replace sigmoidal transition curves by a pair of rectangular step-functions. A first function Y_i switches when a transition starts and prefigures the future state of the variable after a typical delay (E_i for a turn-on, D_i for a turn-off of Y_i). A second function y_i switches when the transition is complete, that is, when the modeled continuous variable passes through a certain threshold. Figure 1 shows the relations between the continuous variables and the Boolean ones.

One supposes (it is a simplifying assumption) that the delay is inertial in the sense that if two opposite transitions occur in a time interval associated with the

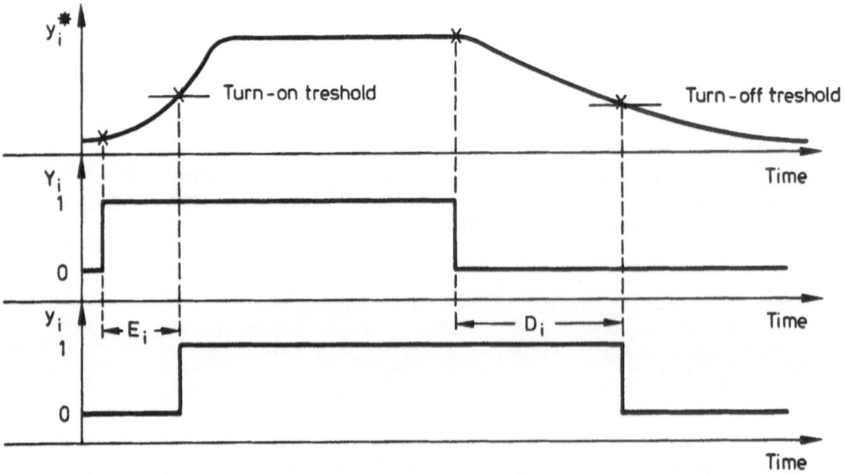

Fig. 1. One may see that Y_i is the input of a "Delay" function and Y_i its output

delay of the first, the two transitions annihilate each other. Figure 2 illustrates this point.

The variables y_i may be viewed as describing the "present" state of the system. Each Y_i representing the "next state function" is naturally a Boolean function of the present state. So, one may say that the kinetical aspects of the system are included in the fact that each state variable transition will take its new value after a specific delay. Figure 3 shows the relations between Y_i and y_i as the well-known asynchronous sequential machine.

It is obvious that when multiple transitions occur, that is, when the next state differs from the present one by more than one bit, the relative values of the delays will decide the first variable which will actually switch. This change in the present state will modify the functions Y_i and certain transitions will be no more excited (annihilated), others will be still excited or newly excited.

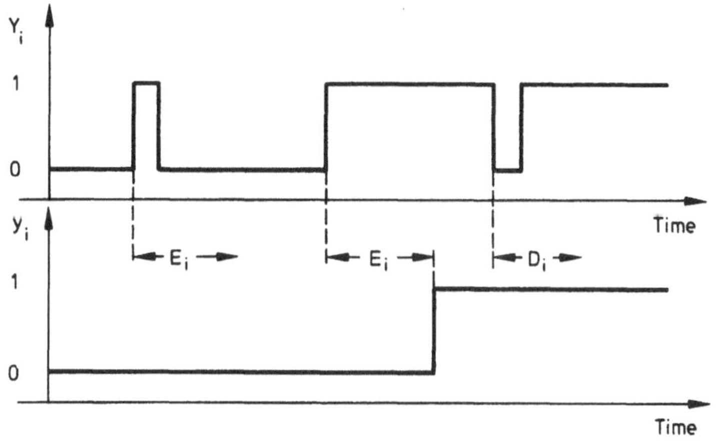

Fig. 2. Two consecutive annihilations in an inertial delay

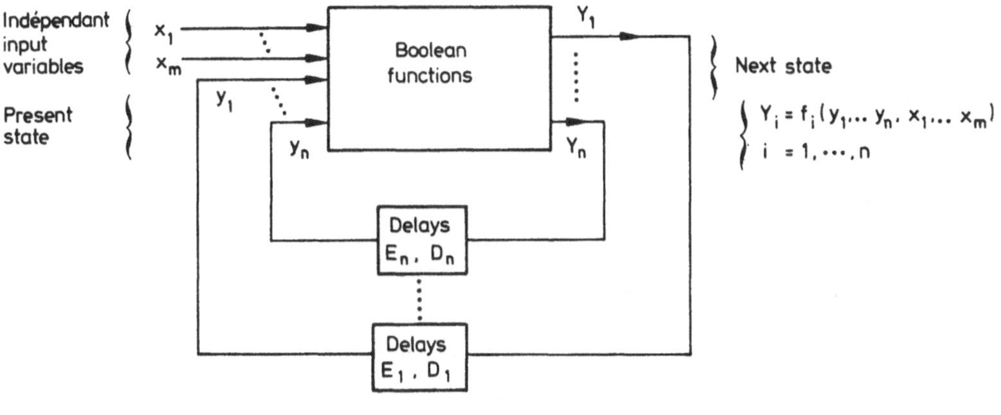

Fig. 3. The classical asynchronous sequential machine structure

This gives the possibility to compute the trajectory of the system in the state space as far as it reaches a "stable state" (next state is equal to present state) or a stable cycle of state. Methods to compute and analyze those trajectories in the case of a great number of variables are described elsewhere [3,4,5,6].

2. Critical Values of the Delays

The model maker is faced with experimental data which, on the one hand, deal with the interactions inside the system and have to be modeled by Boolean functions, and on the other hand are related to the delays. Given the Boolean functions describing the logical structure of a model ($Y_i=f_i(y_1,...,y_n)$; $i=1,...,n$), the modeler knows that any change in the behaviour of the model is due to a modification of the delays. The delays when they are available are generally given by a mean value with dispersion depending on the experimental methods actually used. This gives rise to the first question: given the logical structure of the model, the mean values of the delays, the standard deviations and an initial state, what are the possible final states (or cycle of states) and their frequency of occurrence?

A program, PRAN 2 [3], was made to answer this question. Due to the fact that this method gives no information about the values of the delays related to a given final behaviour, another program PRAN 3 is now available. It computes the mean value of the delays and their variance with respect to each final behaviour computed.

Proceeding by simulation, PRAN 3 gives thus a first rough image of the "delays space" and delays bounds associated with the behaviour. If we adopt the model-maker point of view, this first approach makes it possible to use other analysis tools to a restricted domain of the delay space. In this range of values, the behaviour is practically the same everywhere but certain "pathological" behaviours remain eventually inside the domain and certainly on its bounds. By their construction PRAN 2 and 3 are not able to compute the critical value of the delays responsible for these behavioural modifications. The reason is that the pseudo-random generator of delays changes all the delays at a time for each simulation, the resulting final behaviour being only stored and counted.

3. The Program VALCRI 1

We will still adopt the point of view of a model-maker who is in possession of experimental data. By means of PRAN 3, he has now an idea of the volume of the delay space where the special baheviour he is interested in is present. The program VALCRI 1 makes a systematic scanning of a predetermined volume of the delay space. It is divided into two parts which correspond to different approaches.

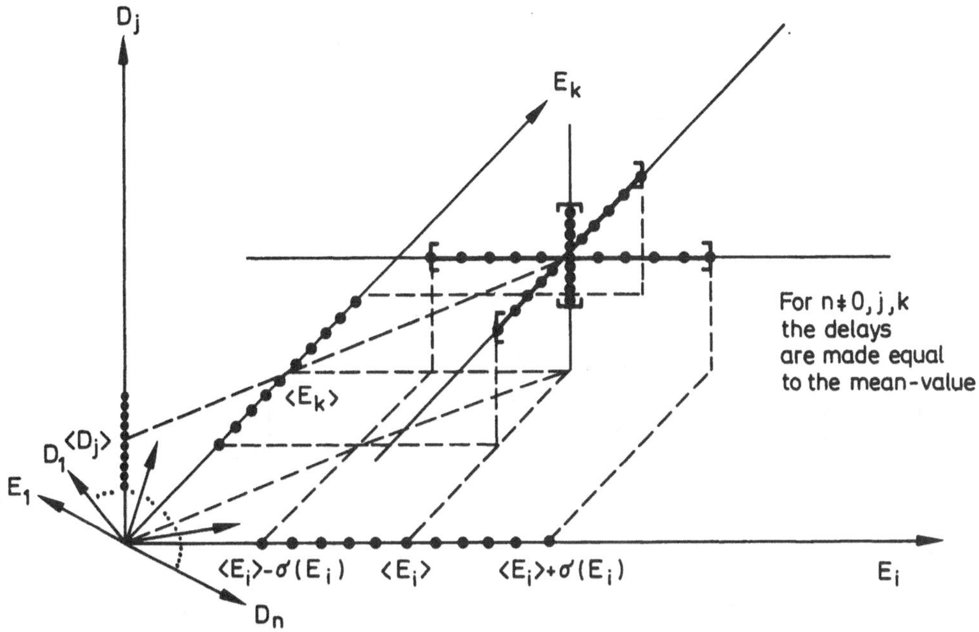

Fig. 4. Shows the first scanning possibility of VALCRI 1

The 2n mean-values given by PRAN 3 (if there are n state Boolean variables) of the delays (E_i, D_i, i=1,...,n) define a point of the delay space, the interesting volume is around this point. The first approach comes from the supposition that the delays are not strongly correlated so that the scanning may be made only along parallels to the coordinate axes. VALCRI 1 makes it possible to perform such an exploration in three directions (for three delays) in one pass. Given the final behaviour for ten equidistant points around the mean and for each of the three chosen delays. Figure 4 illustrates this point. Of course, each time a new behaviour is detected, it is printed with the corresponding delay values.

The second approach consists in a volume exploration, that is, simultaneous variations of two or three delays. This last limitation comes from the fact that we want to give "visual" information to the model-maker. As in the first part, VALCRI 1 makes the nonvariable delays equal to an imposed input value (generally the mean-value given by PRAN 3). The printed results are made of successive layers. Each of them is a rectangular area where the final behaviours computed are indicated in 121 points. Figure 5 shows how the exploration is made.

Obviously in the geometrical layer representation, VALCRI 1 gives the different behaviours by a code which is defined by the program itself and printed elsewhere in the results with recall of the corresponding values of the delays.

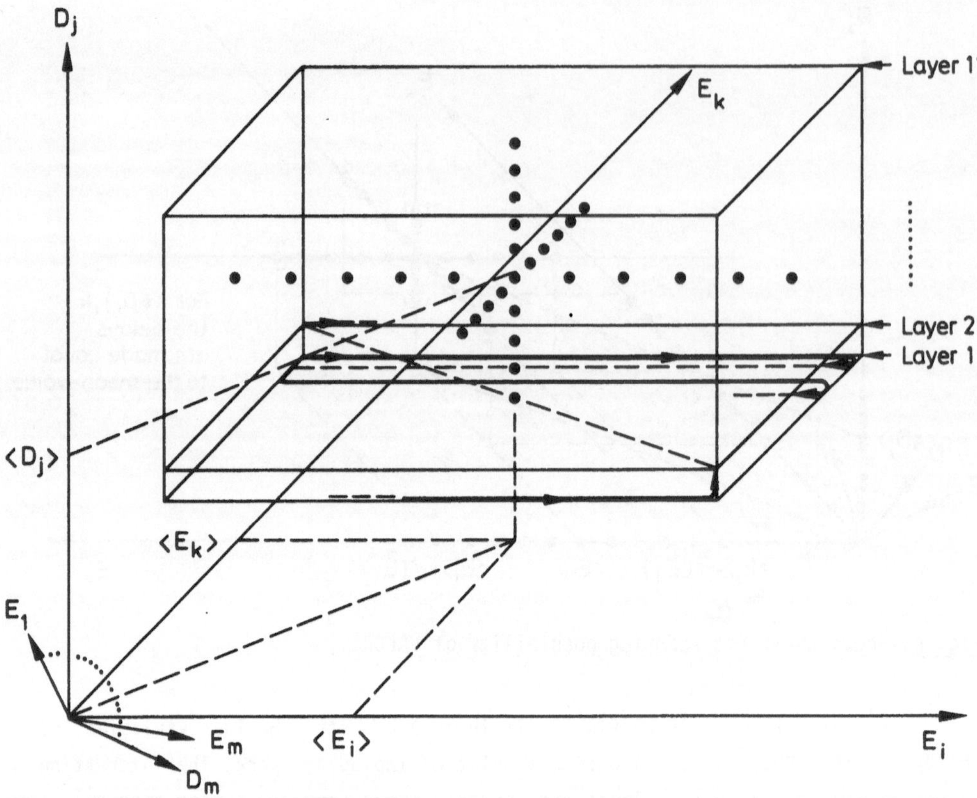

Fig. 5. A succession of 11 layers of 121 points each are scanned and printed layer after layer

4. Theoretical Aspects of the Problem

Programs PRAN 3 and VALCRI 1 are powerful tools for finding the values of the delays which are critical in the sense that they represent bounds implying a drastic modification in the behaviour of the model. Those results are very useful in a model-maker-experiment dialogue. However they do not lead to a deeper understanding of the model. The reason is that the most desirable results may be viewed as a set of linear inequalities with delays as variables and which is under its irreducible and simplest form. This set of inequalities would imply that the system described will always move from the given initial state to the chosen final behaviour. The exact trajectory in the state space is not fixed. As far as we know, there is no simple method to transform the results of VALCRI 1 (which is the description of an hyper-volume in the delay space) into such a set of inequalities.

On the contrary, René THOMAS has developed a very practical method to compute a set of inequalities associated with a pair of initial-final states, independently of any numerical value of the delays. The reduction of this set of inequations to an irreducible one is (again, as far as we know) neither practical nor systematic for great systems.

A program PRFO 3 [4] gives also this set of inequations in the case where it is associated with a precise path in the state space of the system. A test of compatibility is also performed.

In conclusion, at each extremity of the problem a theoretical gap seems to exist before this irreducible set of inequalities is obtained. In this paper, our purpose is twofold:

- first to show that some practical tools of investigation are immediately available by the model-maker (programs are written in FORTRAN IV) even if one has to handle experimental data,
- second to ask if the mentioned theoretical problem does actually exist or not.

Acknowledgement. The authors wish to thank Professor THOMAS for the many fruitful discussions.

References

1. R. Thomas: J. Theor. Biol. *42*, 563 (1973)
2. R. Thomas, P. Van Ham: Biochimie *56*, 1529 (1974)
3. R. Thomas (ed.): *Kinetic Logic*, in Lecture Notes in Biomathematics, Vol.29 (Springer, Berlin, Heidelberg, New York 1979)
4. P. Van Ham: Ph.D. Thesis, University of Brussels (1975)
5. P. Van Ham: Symposium: Applied Aspects of the Automata Theory. Varna, Vol.2 (Edit. Bulgarian Academy of Sciences, Institute of Engineering Cybernetics 1975) p.728
6. P. Van Ham: IFAC Symposium: Discrete Systems, Vol.5 (Dresden, KDT & WGMA 1977) p.27

Chapter 4
Applications in Chemistry

A Simulation Technique for Studying Critical Properties of Chemical Dissipative Systems

P. Hanusse

Centre de Recherche Paul Pascal - CNRS, Domaine Universitaire
F-33405 Talence Cêdex, France

1. Introduction

In this article we present a Monte-Carlo simulation algorithm for studying the behaviour of fluctuations in chemical model systems evolving far from equilibrium. These systems are known as chemical dissipative systems after I. PRIGOGINE and coll. |1-4|. They provide a good example of the variety of phenomena that can be expected and indeed observed when a dynamical system is driven far from equilibrium by an external constraint like an influx of matter or energy or both.

In section 2, we first recall the basic results of the deterministic theory of these systems. Yet, at this level, some analogy with equilibrium phase transitions appears. Next we consider the stochastic theory that is usually used to describe the behaviour of fluctuations in such systems, in particular near bifurcation or transition points where some qualitative change appears in the dynamical properties of the system. Soon, a need for numerical simulations appears. In section 3, we present a Monte-Carlo simulation algorithm specially designed for studying large reaction-diffusion systems within the framework of the stochastic theory mentioned above. Comparison is made with a standard method. Finally in section 4, we discuss results obtained for the Schlögl model in one dimension. Although enhancement of fluctuations and increase in correlation length is observed near the critical point defined by the deterministic theory, no true critical behaviour exists. We compare simulation results with an approximate analytical approach from MALEK-MANSOUR and HOUARD |5|.

2. Chemical Dissipative Systems

2.1 Deterministic approach

Like other systems, chemical dissipative systems are essentially dynamical systems. Their dynamical nature results from usual chemical kinetics. Constraints are achieved either by a controlled influx of some reactants which is usually the case in a typical experimental set-up like a continuous strirred reactor or by considering that the concentration of some species is kept constant which is usually done when studying model systems. In this context, it is clear that one obtains a set of non-linear differential equations that describe the evolution of the concentration of the various intermediate species that are not directly controlled. There exist a number of experimental studies on these systems |4|. But so far none of them has been devoted to the question of fluctuations which is in fact a really hard problem. For this reason, there is quite a distance between real systems and model systems of the kind we shall be concerned with. Here are two examples :

$A \rightarrow X$
$B+X \rightarrow Y+D$ (I)
$2X+Y \rightarrow 3X$
$X \rightarrow E$
(A, B, D and E fixed)

$A \rightarrow X$
$2X \rightarrow 2Y$ (II)
$Y+Z \rightarrow 2Z$
$X+Z \rightarrow B$
(A and B fixed)

Model I is the celebrated "Brussellator" |1| which is a sort of "Ising model" in this field. Model II has been studied by the author |6,7| as a proof that mono and bimolecular steps only could lead to a variety of dissipative structures provided there are at least three independent variables |8|. The various kinds of structures that are to be expected in these systems include limit cycle oscillations, fixed spatial structures |1|, oscillating spatial structures |9|, wave propagation and even chaos |10|. The common feature of all these phenomena is their emergence at a given distance from equilibrium when driving forces are "increased", here the concentration or influx of some controlled species. Let us call B such a bifurcation parameter. Then the dependence of some property like the amplitude of a limit cycle or the value of the stationary concentration on this parameter, may well be described by fig. 1. Clearly, an order parameter can be defined here as in second order equilibrium phase transitions. In other cases the stationary concentration will evolve as shown in fig. 2.

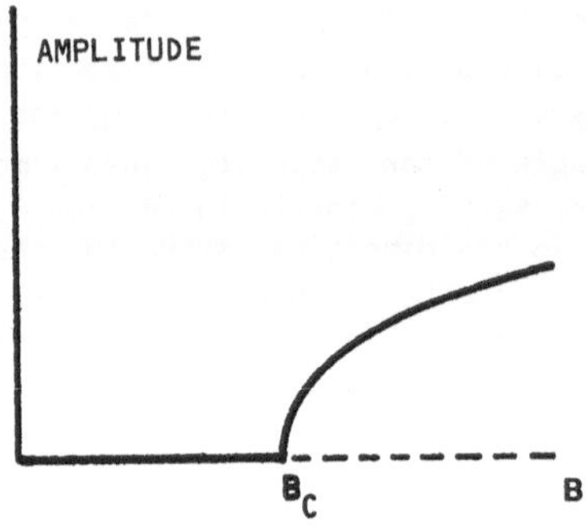

Fig.1 Amplitude of limit cycle versus bifurcation parameter. For low B there exists one stable stationary state. This state becomes unstable beyond B_C. A limit cycle oscillations is then the only stable regime.

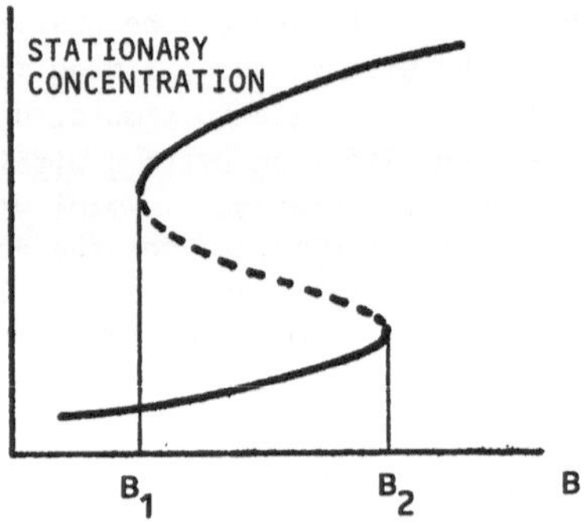

Fig.2 Stationary state versus bifurcation parameter in a bistable system between B_1 and B_2. A critical point exists when B_1 and B_2 meet each other when changing a second constraint parameter.

From this very brief discussion it is clear that there is an at least formal analogy between the behaviour of these dynamical systems and equilibrium phase transition |11-14|. At present investigating the validity of this statement is one of the major concern. It is also why the question of the behaviour of fluctuations arises so naturally. Thus we need some theory that incorporates these fluctuations in the dynamical description itself. This is the role of stochastic theory.

2.2 Stochastic approach

There are different ways one could think of when trying to include the effect of fluctuations. One is to consider that external constraints do not have exactly fixed values but fluctuate about a mean value with some particular distribution. This leads to Langevin type equations or stochastic differential equations. But this is not going to give any information on spontaneous internal fluctuations. It will tell us how the system behaves as a sort of active filter to external noise |15|. This an interesting problem but not the one we address ourselves to. To study internal fluctuations one could in principle start at the microscopic level but it would be a very hard task if not impossible. Besides, there is no evidence that such a detailed theory is required to describe fluctuations of macro-variables. Thus a macroscopic theory is commonly used. Namely a birth and death process description of chemical kinetics, that is a random walk in the particle number space. Let us consider the following simple model known as the Schögl model,that we shall use later on.

$$A + 2X \underset{k_2}{\overset{k_1}{\rightleftarrows}} 3X$$

$$B+X \underset{k_4}{\overset{k_3}{\rightleftarrows}} C$$

(A, B and C fixed)

We take the following values for the various parameters introducing three new control parameters δ, δ' and γ |16| : $k_1A = \frac{3}{A}(1-\gamma)$; $k_2 = \frac{1}{A^2}$; $k_3 = 3+\delta$; $k_4B = A(1+\delta')$; $x = \frac{X}{A}$; $\gamma=0$.

Now the state of the system will be completely specified by the probability $P(X;t)$ for finding X particles of the corresponding independent variable in the system at time t. Each reaction step will change this particle number. For instance, the probability that on particle appears through step $A+2X \rightarrow 3X$ during time internal Δt will be $k_1 A\ X\ (X-1)\ \Delta t$ |17-19|. From this we obtain the master equation that describes the evolution of $P(X,t)$:

$$\frac{\partial}{\partial t}P(X,t) = M(X+1)\ P(X+1;t) - M(X)\ P(X;t) + N(X-1)\ P(X-1;t) - N(X)\ P(X;t)$$

with

$$M(X) = \frac{1}{A^2}\ X\ (X-1)\ (X-2) + (3+\delta)X$$

$$N(X) = \frac{3}{A}\ X\ (X-1) + A(1+\delta')$$

which are death and birth rates or probability per unit time. This is what is called a global description in which only the total number of particles is considered. This is not sufficient and even realistic since local fluctuations are ignored. Thus a local description will include a diffusion process.

The system is partitionned into a set of n identical cells between which particles can diffuse. A multivariate master equation is obtained :

$$\frac{\partial}{\partial t} P(\{X\};t) = \sum_{i=1}^{n} \Big| M(X_i+1)\, P(X_i+1;t) - M(X_i)\, P(\{X\};t)$$
$$+ N(X_i-1)\, P(X_i-1,t) - N(X_i)\, P(\{X\},t)$$
$$+ \frac{D}{2}(X_i+1)\, P(X_i+1, X_{i+1}-1;t) - \frac{D}{2} X_i\, P(\{X\};t)$$
$$+ \frac{D}{2}(X_i+1)\, P(X_{i-1}-1, X_i+1;t) - \frac{D}{2} X_i P(\{X\};t) \Big|$$

where D is a stochastic diffusion coefficient ; only particle numbers different from the generic state {X} have been specified in P. This equation is valid in a one-dimensional system but is straightforwardly extended to higher dimensions.

Clearly the knowledge of P({X};t) allows us to calculate all the moments and in particular the mean concentration <X>, the variance $<\delta X^2>$, with $\delta X = X-<X>$, and correlation functions like the spatial correlation function, $C(l) = <\delta X_i\ \delta X_{i+l}>$ where i denotes the cell number. Stationary properties will be considered only. This is quite enough. As one might expect no exact analytical solution can be usually found even for simple models like the one previously mentioned. This is essentially due to the non-linearity of reaction processes, and also the large number of variables in the local description. Worse than that is that there is no systematic expansion procedure that works near bifurcation points. Only far from them does a volume expansion |18| or moment |20| or cumulant expansion |14| work. When applied to a multivariate master equation they use a mean-field approximation that fails near instability points |21,22|. So, we are somehow at a loss ! Recently MALER-MANSOUR and HOUARD |5| have proposed a new approximate theory -MH sheme- that goes beyond the mean field assumption. Our aim is in fact to investigate the validity of their assumption. It seems to be working fairly well at least in one dimensional systems. They assume that the conditional mean value in cell i+1 knowing the state of cell i is given by

$$<X_{i+1}>_{X_i} = <X> + (X_i-<X>)\ \frac{<\delta X_i\ \delta X_{i+1}>}{<\delta X_i{}^2>} \quad .$$

Of course there are arguments leading to this form but no systematic perturbation approach of some kind is used. Clearly the only way to check this approach is to obtain an exact numerical solution of the master equation, which does not mean calculating P({X},t) explicitly, but only relevant moments, in particular the spatial correlation function.

3. Monte-Carlo Simulation Methods

A Monte-Carlo simulation method is exactly what we need. It has been used in various fields to study equivalent problems |23|. Let us mention the dynamic Ising model for which a birth or death description can be used |19|. We have been using various standard or specific formulations for some time, essentially to obtain qualitative results |24,7,9|. Other authors have been doing the same sort of thing |25,26|. We have recently proposed an algorithm |27| based on earlier studies that is particularly designed to study large reaction-diffusion systems. Before discussing it -and presenting some results in the next section- we recall the standard method that is called the "minimal process method". As we shall see it is not that "minimal" in the sense that it is an n^2-algorithm where n is the number of cells in the system.

The theoretical basis of this method results from the markovian character of the birth and death dynamics. It can be demonstrated |28| that for any markovian process the waiting time is distributed exponentially. Let W be the waiting time in the current state, then one has :

$$\text{Prob}\,(W>t) = \exp(-V\,t)$$

where V is the sum of the rates of all processes, birth, death and diffusion, in all cells. The system being in a given state we can determine randomly the time interval Δt the system will stay in this state which is obtained by $\Delta t = \frac{1}{V}\,\text{Log}\,T$, where T is a uniformly distributed random number in $|0,1|$. The process that will occur next will be randomly picked up among N possible ones having rate v_i, with a probability v_i/V. Notice that in the kind of system we are interested in several hundreds of processes can occur at a given time. We can use a set of variables C_i defined by $C_i = \sum_{j=1}^{i} v_j, i=1,2 \ldots N$. They represent cumulated rates and from this definition one can see that C_N is equal to V. This array variable will enable us to select randomly a process by a simple binary search once one has picked up a random number between 0 and V, which is a fast way to find the process number k. From this number, the cell number and the process type in the cell, are easily determined. After the process is performed we do not have to recalculate all the C's but only those corresponding to the modified cells and to shift those to the "right" of these cells by an fixed amount easily determined. In this procedure, the most time consuming steps, updating the C's and picking up a process are both n-dependent. For this reason the calculation time will be roughly proportional to the square of the cell number n for a given model system.

Let us now present a more effecient algorithm. As said before, in the minimal process algorithm most of the time is spent in finding out which process will occur among the N ones and in updating the C's. In the following algorithm we find out first where, i.e. in which cell, the next process is going to occur and then which process among the p ones, a quite small number, is effectively occurring in the selected cell. This is made possible in an efficient way by introducing an extra process that we call a "null process", the rate of which is calculated for each cell in such a way that the total rate for each cell including the null process is equal to the same quantity that we call Q, independent of the cell number. Introducing this null process allows us a further simplification. After performing each process including the null one, the time t is incremented by an amount $1/nQ$, where n is, as before, the number of cells. This amounts to taking systematically an average waiting time after each process, since nQ is the total rate, equivalent to V in the previous section. But this is not to say that the system will always stay in given state for this time interval, since several null process may occur before an effective state changing process occurs. Besides Q is not constant. It is continuously adjusted to the total rate of the "largest" cell, which is of course the cell where the probability that a process occuring next is the highest. Adjusting Q in this way minimizes the number of null processes which is specially necessary when they are large local fluctuations. Notice that the non-constant time step $1/nQ$ is much smaller than the characteristic time of any real process is the system. A theoretical justification of this procedure is given in |27|.

Our procedure is thus very simple. Once the cell number has been determined by picking up a random number between 1 and n, we can quickly find out which process will occur among the p ones. In the simplest system p is equal to four with a one-dimensional system and six in a two-dimensional system. Clearly the time consumed for this search is independent of the cell number n. So we see that performing null processes does not mean necessarily wasting time. To find out as soon as possible, that is with the minimum amount of calculation, that a null process has been picked up, we now define C_i as the total effective rate in cell i. Having picked up a random number T, $0 \leq T < Q$, a null process occurs if T is greater than C_i. At least for systems with one species, processes rates like M(X) or N(X) and the values of C(X), for a number of particles equal to X, need be calculated only once at program initialization level and stored in an array variable. In this way very little calculation is done inside the inner loop of the program. This is not possible for the C's used in the algorithm described before. A flowchart for this algorithm is given in fig.3.

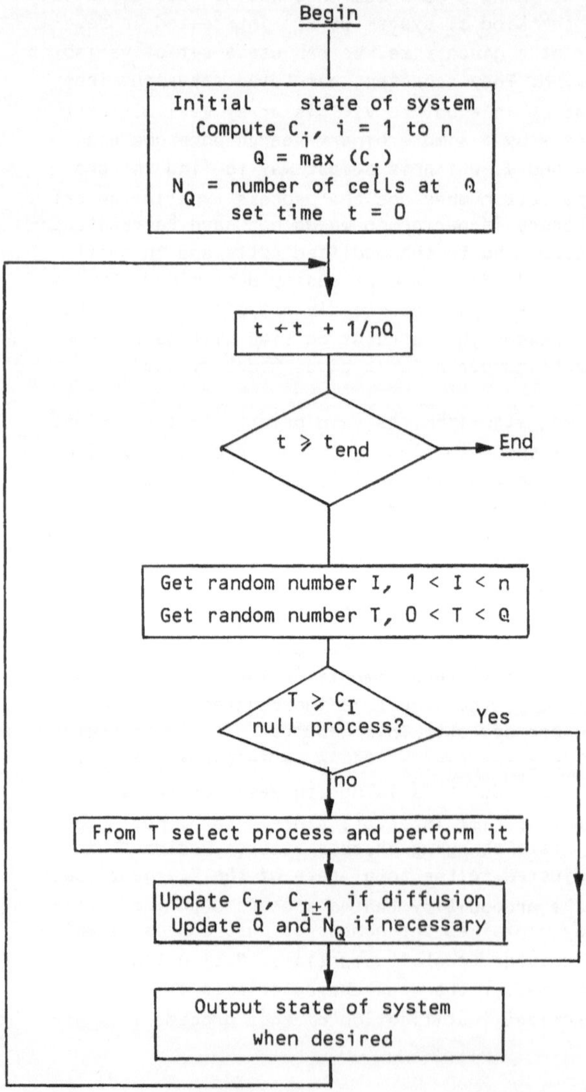

Fig.3 Flowchart of Monte Carlo procedure

We have performed extensive tests to compare these two algorithms. As expected the latter which is n-dependent is much faster than the standard one which is n^2-dependent. Typically by factor 8 with about 100 cells. The standard method has another disandvantage, namely, its sensitivity to random number generator imperfections. More details will be found in ref. 27, along with a survey of various problems encountered when using this kind of technique. We have calculated the spatial correlation function of a simple model system for which an analytical solution is available |29|. Extremely good results have been obtained even for long range and/or small correlation. We are now equipped to investigate a real interesting problem.

4. The Critical Point Of The Schlögl Model

The deterministic kinetic equation of this model (see section 2.2) is given by

$$\frac{dx}{dt} = -(x-1)^3 - 3\gamma x^2 - \delta x + \delta' .$$

There is a triple point for $\gamma = \delta = \delta' = 0$. Considering the parameter subspace $\gamma=0$, $\delta=\delta'$ we see that for $\delta>0$ there is one unique stable stationary state at $x=1$ and for $\delta<0$ this state is unstable and two other stable states $x=1 \pm \sqrt{-\delta}$ appear. The stationary states dependence on δ is looking very much like fig. 1 to which a symetric negative branch would be added |16|. This is strongly reminiscent of the critical point of the Van der Waals equation. What is the stochastic equivalent of this ? Incidently, it should be pointed out that the position of the critical point in the (δ,δ',γ) space may be defined in different ways, all becoming identical in the thermodynamic limit corresponding here to $A \to \infty$. For instance one could define the critical point in the stochastic for formulation in the triple point of $M(X) - N(X) = 0$, which gives $\gamma=1/A$; $\delta = -(3A-1)/A^2$; $\delta'=0$. In the sequel we shall refer to the usual definition and restrict ourselves to $\gamma=0$; $\delta=\delta'$.

Now we are facing two separate questions. First, how can we calculate analytically the spatial correlation function when usual mean-field type approximations fail, that is when fluctuations have large amplitude ? Second, does the Schlögl model present true critical behaviour ?

The MH scheme yield the following equation for the spatial correlation function in a n-cell one dimensional periodic system.

$$C(|k-l|) = \langle\delta X_k \delta X_l\rangle = \langle X\rangle\, \delta^{Kr}_{kl} + \frac{bR}{D(R^2-1)(R^n-1)} \left[R^{|k-l|} + R^{n-|k-l|}\right]$$

with

$$R = 1 + \frac{c}{D} + \left(\frac{c}{D}\left(\frac{c}{D}+2\right)\right)^{1/2}$$

$$c = \frac{\langle X(M(X)-N(X))\rangle}{\langle \delta X^2\rangle}$$

$$b = 2\langle X\rangle \left(\frac{\langle M(X)\rangle}{\langle X\rangle} - c\right)$$

The correlation length is seen to be $L_C = -\frac{1}{\ln R}$. All the quantities appearing above can be calculated numerically in a self consistent way in the MH scheme.

We have performed Monte-Carlo (MC) simulations for various values of δ and for various system sizes n. In table I we present a few results at $\delta=0$ for n = 15, 21, 31, 41 along with the MH estimates. Other results obtained at fixed n and various $\delta>0$ show that the MH estimates are closer to MC results the farther one stays from $\delta=0$. The data in table I also show that the smaller local fluctuations ($\langle \delta X^2 \rangle$) the better the agreement. But most important is that the MH scheme gives much better results than any other approximate theory |5|. On the other hand both MH estimates and MC results do not show any critical behaviour.

Table 1 Mean, variance and spatial correlation function of the Schlögl model in one dimension for various system size n for $\delta=\delta'=\gamma=0$; A=20 ; D=20. MC means Monte Carlo results with 90% confidence interval MH means Malde-Mansour, Howard estimates. Periodic boundary conditions

	n=15		n=21		n=31		n=41	
	MC	MH	MC	MH	MC	MH	MC	MH
$\langle X \rangle$	19.05	18.85	18.90	18.90	18.98	18.94	18.93	18.94
+-	.07		.05		.05		.05	
$\langle \delta X^2 \rangle$	51.1	47.6	47.2	45.8	46.4	44.7	46.2	44.4
C(1)	28.5	25.4	25.0	23.4	24.0	22.3	23.9	22.1
2	25.6	22.7	22.3	20.6	20.9	19.4	21.1	19.1
3	23.7	20.6	19.8	18.2	18.2	16.9	18.5	16.6
4	22.0	18.9	17.8	16.2	16.4	14.7	16.3	14.4
5	20.8	17.7	16.2	14.6	14.2	12.9	14.4	12.5
6	19.9	16.9	15.0	13.3	12.7	11.3	12.8	10.8
7	19.5	16.5	13.9	12.3	11.2	10.0	11.2	9.4
8			13.0	11.6	10.1	8.8	10.0	8.2
9			12.5	11.1	9.0	7.9	8.8	7.2
10			12.4	10.3	8.2	7.1	7.7	6.3
11					7.5	6.5	6.8	5.5
12					6.8	6.0	6.1	4.9
13					6.4	5.7	5.3	4.3
14					6.0	5.5	4.9	3.8
15					5.9	5.3	4.6	3.5
16							4.0	3.2
17							3.7	2.9
18							3.6	2.8
19							3.5	2.7
+-	0.7		0.5		0.5		0.4	

Indeed the correlation length no longer increases for systems over 31 cells. The local regime is no longer dependent on system size which can be seen in table I looking at the first few values of the correlations function. This is also confirmed by calculations on larger systems not reported here. Besides when fitting the MC results with the MH function, that is a double exponential, one obtains even better results than MH estimates, although still slightly outside of confidence intervals.

5. Conclusion

We have presented a Monte-Carlo method particularly designed to calculate spatial correlation functions in reaction-diffusion systems. Results obtained for an exactly solvable system make us confident about the results one could obtain by this method. When applied to the Schlögl model near the "singular" if not critical point defined by the deterministic theory this method gives stable and consistent results. These results show that the MH scheme leads to qualitatively correct results and quantitatively fairly good results although the limit of validity of this scheme appears specially when local fluctuations have large amplitude. No true critical behaviour, i.e. continuous increase of correlation length with system size, is observed. This result could appear obvious from the point of view of equilibrium phase transitions where it is well-known that no second order phase transition is possible in one dimension. But care must taken not to take for granted the results obtained there to the problem studied here. Complete analogy remains still be founded on a coherent theory.

Of course, the next step is to study the effect of dimensionality. Preliminary results in two dimensions tends to suggest that MH estimates might give qualitatively incorrect results in this case and critical behaviour is not ruled out at present stage of investigations.

Part of this work has been done with A. BLANCHÉ.

References

[1] G. Nicolis and I. Prigogine, Self-organisation in non equilibrium systems, Wiley, New York (1977)
[2] H.Haken, Synergetics, An Introduction, 2nd ed., Springer Series in Synergetics, Vol.1 (Springer, Berlin, Heidelberg, New York 1978)
[3] Adv. Chem. Physics, vol. 38, S.A. Rice Ed., Wiley New York (1978)
[4] Synergetics, Far from equilibrium, ed. by A. Pacault and C. Vidal, Springer Series in Synergetics, Vol. 3 (Springer, Berlin, Heidelberg, New York 1979)
[5] M. Malek-Mansour, J. Houard, Physics Letters 70 A, 5, 366 (1979)
[6] P. Hanusse and A. Pacault, Proceedings of the 25th International Meeting of the Societé de Chimie Physique, Dijon 1974, Elsevier (1975)
[7] P. Hanusse, Phd. Thesis, University of Bordeaux I, January 1976
[8] P. Hanusse, C.R. Acad. Sci Paris, 274 C, 1245 (1972) and 277C, 263 (1973)
[9] P. Hanusse, in ref. 4 p. 70-74
[10] J.C. Roux, A. Rossi, S. Bachelart, C. Vidal, Physics Letters, to appear
[11] R.J. Mc Neil and D.F. Walls, J. Stat. Phys. 10,6, 439 (1974)
[12] A. Nitzan, P. Ortoleva, J. Deutch, J. Ross, J. Chem. Phys. 61, 3, 1056 (1974)
[13] I. Prigogine, R. Lefever, J.S. Turner, J.W. Turner, Physics Letters 51A, 6, 317 (1975)
[14] H. Lemarchand and G. Nicolis, Physica 82A, 521 : 452 (1976)
[15] W. Horsthemke and R. Lefever, Physics Letters 64A, 1, 19-21 (1977)
[16] G. Nicolis and R. Lefever, Physics Letters, 62A, 7, 469-71 (1977)
[17] D. Mc Quarrie, Adv. Chem. Phys. vol. 15 (1969)
[18] N.G. Van Kampen, Adv. Chem. Phys. Vol. 34, 245 (1976)
[19] I. Oppenheim, R.E. Shuler, G.M. Weiss, Stochastic Processes in Chemical Physics MIT Press (1977)

[20] P. Hanusse, Physics Letters 59A, 421 (1977)
[21] G. Nicolis and I. Prigogine, Proc. Nat. Acad. Sci. USA, vol. 38, 9, 2102 : 2107 (1971)
[22] M. Malek-Mansour and G. Nicolis, J. Stat. Phys. 13,3, 197 : 217 (1975)
[23] K. Binder Ed. Monte Carlo Methods in Statistical Physics, Springer Verlag, Berlin (1979)
[24] P. Hanusse, J. Chem. Phys. 67, 1282 (1977)
[25] D. Gillespie, J. Comp. Phys. 22, 403 (1976)
[26] J.S. Turner, J. Phys. Chem. 81, 2375 (1977)
[27] P. Hanusse and A. Blanché, J. Chem. Phys., to appear
[28] S. Karlin and H.M. Taylor, a first course in stochastic processes, Academic Press , New York-London (1975)
[29] J. Houard, Mémoire de Licence, Université Libre de Bruxelles (1978)

Critical Paths and Passes: Application to Quantum Chemistry

D. Liotard* and J.-P. Penot**

(*Département de Chimie) (**Département de Mathématiques)
Université de Pau, Faculté des Sciences, Avenue Philippon
F-64000 Pau Cédex, France

The purpose of this paper is to describe a process for finding a critical point of a function f and a path connecting this critical point to two given points (which are usually local minima of the function). This problem issues from quantum chemistry : the function f represents the energy of a molecule and, given two local minima of f (which correspond to stable molecular states), one looks for a "reaction path" connecting them. Such a path is required to make the variation of f along it the least possible; the highest point on the path being a saddle point (or pass) whose knowledge is of utmost importance reaction rates theory [1].

The organisation of this paper is as follows : the first part deals with the problem of finding some more critical points ; the second part describes a practical algorithm for locating such points and a path connecting them. Although we do not have a proof of convergence of this algorithm, some mathematical foundations are given at the end of part one for nice functions (Morse functions). Numerical experiments and further theoretical developments will be published elsewhere. A more global point of view is adopted in [2] where Morse theory is invoked for discovering new critical points and applied to the same chemical problem.

1. Mathematical Foundations

In this section we present some theoretical results from global analysis useful for finding critical points of a C^1-function on a domain X of a Banach space E. We set $d(x) = \max(\|x\|, d(x, E\setminus X))$ for $x \in X$ and we denote by K the set of critical points of f on X, $X_r = f^{-1}((-\infty, r])$ for $r \in \mathbb{R}$.

1.1 Finding more Critical Points

The following result is a variant of [3].

Proposition 1 :

Let us suppose :

(H_0) *E is finite dimensional ;*

(H_1) *There exists x_0, x_1 in X, an open subset B of X containing x_0 but not x_1, with boundary $S \subset X$ and $\alpha \in \mathbb{R}$ such that $f(x_0) < \alpha$, $f(x_1) < \alpha$ and $f(x) \geq \alpha$ for each $x \in S$.*

(H_2) *$\alpha < \ell = \liminf_{d(x) \to +\infty} f(x)$ and there exists a connected subset C of X containing x_o and x_1 with $\sup f(C) < \ell$.*

Then, if, for $r \geq f(x_o)$, C_r is the connected component of x_o in X_r

$$h = \inf \{r \mid x_1 \in C_r\}$$

is a critical value of f and $h \in [\alpha, \ell]$.

The proof, which uses only elementary arguments, will not be given here. Let us note that this result can be illuminated by the following picture : the altitude of a pass between two valleys is the least height at which two separated horizontal fog-banks in each valley can be joined into a unique set. However it is of little practical value in determining the level sets X_r and their connected components would be too costly, in particular when $\dim E > 2$.

Corollary 2 :

If f is coercive (i.e. $\lim_{d(x) \to +\infty} f(x) = +\infty$) and if f has a local strict minimum x_o which is not a global strict minimum then f has another critical point.

Proof :

As f is not a strict global minimum, there exists $x_1 \in X$ with $f(x_1) < f(x_o)$. Then B can be chosen to be a ball with center x_o and radius $\rho < d(x_o, x_1)$ small enough so that $f(x) > f(x_o)$ for each $x \in \bar{B}$. □

The method of proof of [3a], which uses the fact that the space of compact connected subsets of a compact metric space is compact when endowed with the Hausdorff distance yields the following more precise result.

Proposition 3 :

Under assumptions (H_o), (H_1), (H_2) the family $\mathbb{C}$ of compact connected subsets of X containing x_o and x_1 has a member $\bar{C}$ such that

$$k = \inf_{C \in \mathbb{C}} \max f(C) = \max f(\bar{C})$$

is a critical value of f and $\bar{C}$ contains a critical point $\bar{x}$ with $f(\bar{x}) = k$.

This can be extended to the infinite dimensional case in the spirit of [4] th. 2.4 with the help of the following Palais-Smale type condition

(H_2') There exists $\ell \in (\alpha, +\infty)$ and $m \in (-\infty, \alpha)$ such that each sequence (x_n) in $f^{-1}([m, \ell])$ with $\lim f'(x_n) = 0$ has a converging subsequence. Moreover, for each $r \in (m, \ell)$ X_r is closed in E.

(H_2'') There exists a connected subsed C of X with $\sup f(C) < \ell$, $x_o \in C$, $x_1 \in C$.

Proposition 4 :

Suppose assumptions $(H_1),(H_2'),(H_2'')$ *hold and let* $\mathfrak{D}$ *be the family of closed subsets of* $X\setminus\{x_1\}$ *containing* x_o *in their interior* $\mathring{D}$ *. Then*

$$d = \sup_{D\in\mathfrak{D}} \inf f(D\setminus\mathring{D})$$

is a critical value of f *.*

This result is dual to the following one (adapted from [4] th. 2.1) which is closer to our aim.

Proposition 5 :

Suppose $(H_1),(H_2')$ *and the following assumption hold*

(H_2''') *There exists* $g_o \in G = \{g \in C([0,1],X)\ ,\ g(0) = x_o\ ,\ g(1) = x_1\}$ *with* $\sup_{0\le t\le 1} f(g_o(t)) < \ell$ *.*

Then

$$b = \inf_{g\in G}\ \sup_{0\le t\le 1} f(g(t))$$

is a critical value of f *with* $\alpha \le d \le b < \ell$*.*

However we do not know if there is some $g \in G$ with $b = \sup f(g([0,1]))$ and a critical point on $g([0,1])$; such a path might be called a critical path.

1.2 Finding a Critical Path and a Pass

In this subsection E is finite dimensional and f is a C^2 *Morse function* (i.e. all its critical points are non degenerated). Let us call a critical point a of f a *pass* if $f''(a)$ has exactly one negative eigenvalue (i.e. if the index ind (f,a) of f at a is one). Let us recall that a *Morse chart* for f at a is a C^1-diffeomorphism $\varphi : U \to V$ where U is an open subset of X containing a , V is an open ball in E with center 0 , $\varphi(a) = 0$ and

$$f(\varphi^{-1}(x)) = f(a) - (x_1^2 + \dots + x_i^2) + (x_{i+1}^2 + \dots + x_d^2)$$

with $d = \dim E$, $i = \text{ind}\,(f,a)$. If f is a Morse function, it has a Morse chart at each of its critical points (Morse's lemma).

Lemma 6 :

If (φ,U,V) *is a Morse chart for* f *as above,* $W = \{x \in U \mid f(x) < f(a)\}$ *is arwise connected if* $i \ge 2$ *.*

This follows easily from the fact that any two points of $\varphi(W)$ can be joined by an arc composed of one or two segments and an arc of circle along which f is constant. Another arc is described in the pioneering work [5] which contains the germ of the idea of the following result, as noticed by the first author.

Theorem 7 :

If f is a Morse function on X satisfying (H_1) and (H_2) then x_o and x_1 can be joined by a path $g \in C([0,1],X)$ whose image contains a pass $\bar{x}$ with $f(\bar{x}) = k = \sup_{0 \le t \le 1} f(g(t))$.

In fact g can be chosen in $C^1([0,1],X)$ and such that $f^{-1}(k) \cap g([0,1])$ is made of a finite number of passes.

Proof :

There is no loss of generality in assuming the compact connected set $\bar{C}$ of proposition 3 satisfies $\bar{C} \cap f^{-1}(k) \subset K$. In fact $\bar{C}$ can be replaced by $\eta_1(\bar{C})$ where η is the flow of a C^1-vector field on X which coïncides with $-\nabla f$ on $f^{-1}([\alpha,\ell])$ and is zero at x_o and x_1 . As the critical points of f are isolated, $\bar{C} \cap f^{-1}(k)$ is finite : $\bar{C} \cap f^{-1}(k) = \{x_2,\ldots,x_n\}$. For $j = 2,\ldots,n$, let $\varphi_j : U_j \to V_j$ be a Morse chart at x_j . By compactness

$$\varepsilon = k - \sup \{f(x) \mid x \in \bar{C} \setminus U_1 \setminus \ldots \setminus U_n\} > 0 .$$

As f is uniformly continuous on the compact set X_k , we can find $\delta > 0$ such that $|f(x) - f(y)| < \varepsilon$ for $d(x,y) < \delta$, $\inf \{d(x,E \setminus X) \mid x \in \bar{C}\} > \delta$, and U_j contains the closed ball with center x_j and radius δ for $j = 2,\ldots,n$. As $\bar{C}$ is compact and connected, we can find a finite sequence $\{y_1,\ldots,y_s\}$ containing $\{x_o,x_1,\ldots,x_n\}$, with $y_1 = x_o$, $y_s = x_1$ and $d(y_i,y_{i-1}) < \delta$ for $i = 1,\ldots,s$. If y_i or y_{i-1} lies outside of the U_j's , then the segment $[y_{i-},y_i]$ is included in X and in fact in $f^{-1}((-\infty,k))$. Let $j \in \{2,\ldots,n\}$ and let y_p (resp. y_q) be the first (resp. the last) term of the sequence (y_i) in U_j . If x_j is of index $i \ge 2$ we replace the broken line $[y_p,y_{p+1}] \cup \ldots \cup [y_{q-1},y_q]$ by an arc in $U_j \cap f^{-1}((-\infty,k))$ using lemma 6. If x_j is of index one, we replace it by the arc $\varphi^{-1}([\varphi(y_p),0]) \cup \varphi^{-1}([0,\varphi(y_q)])$ which lies in $f^{-1}((-\infty,k)) \cup \{x_j\} \subset X_k$. If every x_j was of index at least two we would obtain by such successive replacements an arc joining x_o to x_1 in $f^{-1}((-\infty,k))$, a contradiction with the definition of k . Hence at least one point x_j is a pass and we end with a path passing through the passes of $f^{-1}(k) \cap \bar{C}$. □

As the set of Morse functions is generic in the set of C^2-functions on X , our result holds for "nearly all" C^2-functions.

2. A Practical Algorithm

In this section we describe an algorithm for finding a saddle point S between two minima M_1,M_2 of a C^2-functional f on $\mathbb{R}^N$ and an approximation to a a path joining M_1,M_2 through S . According to practical considerations, the algorithm must only require the explicit knowledge of f and its gradient ∇f at a given point p (the direct evaluation of the hessian is too costly for most chemical applications).

A continuous shifting of a continuous path is unrealistic from the numerical point of view. Therefore a path will be approximated at the ith-iteration by a chain of points : $C_i = \{p_1 \ldots p_{ni}\}$ running from the starting minimum M_1 (the chemical reactants) to the ending one M_2 (chemical products). An iteration will roughly consist in replacing p_H , the highest point of the chain, by a new one p^* according to :

$$p^* = p_H + \lambda \, D \, / \, \|D\| \tag{1}$$

with $\lambda \in [\ell_o, \ell_2]$ and D a shifting vector. The threshold ℓ_0 must be strictly positive. A natural choice for D would be the opposite gradient at p_H : $G = -\nabla f(p_H)$. Numerical experiences suggest the more efficient direction provided by the projected gradient G^* :

$$G^* = G - (G,T)\, T \tag{2}$$

with T the "tangent" to the path at p_H and (G,T) the scalar product between the vectors G and T . We define the tangent as the unit vector parallel to $A_H + A_{H+1}$, where $A_i = (p_i - p_{i-1}) \, / \, \|p_i - p_{i-1}\|$. Some other choices for D will be described later on. We require :

R_1 : each new link of the chain to have a length ℓ with $\ell \in [\ell_1, \ell_2]$

R_2 : the thresholds ℓ_1, ℓ_2 to decrease slowly with increasing iterations.

Successive shifting could generate a rather complicated path with inefficient meanders. This pitfall is avoided by a short analysis of distances between the new point p^* and old links remaining on the left (resp. right) side of p^* as shown on fig. 1.

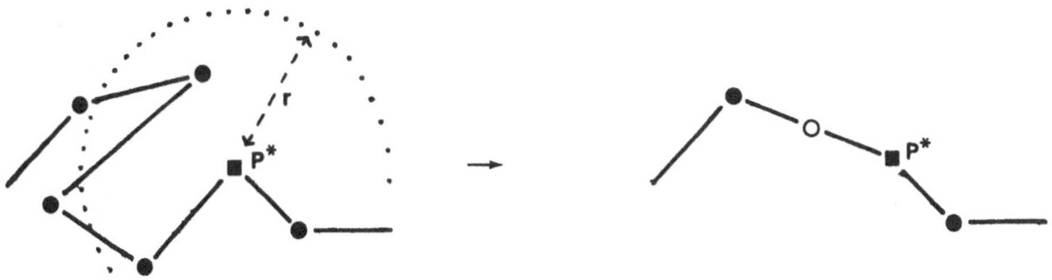

Fig. 1 : meander by-pass

The left chain is cut-off if a (left) link lies inside the sphere centered on p^* with radius r , requiring $r \in [0, \ell_2]$. If necessary, a new point (O on fig. 1) is inserted in the middle of the new link to ensure R_1 .

2.1 Drawing a Quadratic Estimate

Each iteration requires the evaluation of $-G_i$, the gradient vector of the function f at p_H , with P_i the coordinates vector. This cumulative information at successive points surrounding the saddle point S can be used to estimate and update the hessian H in S according to

$$(G_i - G_{i+1}) = H(P_{i+1} - P_i) \ .$$

Taking into account the tendancy of most variable metric methods to accumulate rounding-off errors [6] we estimate directly H using the last $(N+1)$ gradient evaluations in N independant directions. The "linear dependance" threshold is established as suggested in [7]. If required, an extra gradient evaluation is carried out for updating H. Diagonalization of the symmetric H matrix leads to its index and to the quadratic direction Q :

$$Q = H^{-1} G \ . \tag{3}$$

Let q be the euclidian norm of Q and let i_H be the index of H. We say the direction Q is "operative" if the three following conditions are satisfied :

(i) $i_H = 1$ (characterizing a saddle point)
(ii) $q \leq \ell_2$ (avoiding quasi singular hessian)
(iii) $|(Q,G)| > \varepsilon \, q \cdot g$ (numerical stability)
with $g = ||G||$ and ε a positive threshold.

2.2 An Outline of the Algorithm

1. Set the thresholds. Generate a trial chain, accounting for M_1, M_2, ℓ_2.
2. Find p_H. Evaluate $G = -\nabla f(p_H)$. Update H and compute Q (eq. 3).
3. If $||G|| < g_c$ then stop. At the preset threshold g_c, the process is achieved.
4. Select the direction D (increasing interpolation or decreasing shifting, see below).
5. Search optimum $\lambda \in [\ell_o, \ell_2]$. Get the new point p^* (eq. 1).
6. Replace p_H by p^*. If necessary, by-pass meanders and insert new points (request R_1).
7. Decrease ℓ_2 and related thresholds. Return to 2.

A carefull optimization of λ is not essential. The choice will depend on the relative cost of a function evaluation and its gradient. Decreasing ℓ_2 may be done according to various laws. Typically, with i the number of the iteration :
$\ell_2 \text{ (new)} = \ell_2 \text{ (old)} \, [i / (i+1)]$.

2.3 Selecting the Direction D

The choice of D drastically influences the performance of the algorithm. It is essential to make use of the direction Q as soon as operative, thus introducing quadratic termination properties.
Briefly, D will be one of the following directions :

- G^* , the projected gradient (eq. 2),
- Q , the quadratic direction (eq. 3),
- A , one of the two unit vectors of the links adjacent to p_H .

Let g_1 be a pre-convergence threshold on the gradient norm. If Q is operative and $\|G\| < g_1$ then we select the direction Q. This choice induces to perform a decreasing shifting or an increasing interpolation with respect to the sign of (Q,G).

In other cases, the type of the modification is selected first, accounting for the angle θ between G and T, the "tangent" to the path. If $\theta > \eta$ (another threshold) a decreasing shifting will be made. If not, the point p_H is a poor approximation of the "true" highest point of the path and an increasing interpolation is to be done.

Decreasing direction D : If Q is operative and decreasing, it is used in place of the projected gradient recommended by eq. 2.

Increasing direction D : If Q is not operative and increasing, we select the more increasing vector A. In some unusual cases no increasing direction can be found, due to a sharp angle between the two unit vectors : the "tangent" T is meaningless. In such a case, the point p_H is replaced by the middle of the link joining the two points adjacent to p_H in order to restore a regular path.

2.4 An Illustrative Example

As a vizualisable test we generate a two dimensional function as follows :

$$f(x,y) = 250 \sum_{i=1}^{4} (1 - \exp(- e_i\ t_i))^2$$

with $e_1 = 0.5 \quad e_2 = 1 \quad e_3 = 0.3 \quad e_4 = 0.2$

and

$$\left.\begin{aligned} t_1 &= (x - 1) \cos \theta_1 - (y - 6) \sin \theta_1 \\ t_2 &= (x - 1) \sin \theta_1 + (y - 6) \cos \theta_1 \end{aligned}\right\} \theta_1 = 86°$$

$$\left.\begin{aligned} t_3 &= (x - 8) \cos \theta_2 - (y - 1) \sin \theta_2 \\ t_4 &= (x - 8) \sin \theta_2 + (y - 1) \cos \theta_2 \end{aligned}\right\} \theta_2 = - 80° .$$

Such functions simulate very well the behavior of effective chemical potential surfaces. As shown in Fig. 2, three minima (M_1 to M_3) are connected by two saddle points (X_1,X_2). Note that $f(X_2) < f(X_1)$. Three inflexion points (I_1 to I_3) associated with a local non-zero minimum of the gradient norm are also present. They provide a very powerfull pitfall for most gradient norm minimization techniques [7,8].

Let M_3,M_2 be the respective left and right minima. The trial path includes the point $(x = 8, y = 2)$, a very bad approximation of the saddle point X_2. Moreover the point X_1 where a continuous path would run over twice provides a possible artefact of the method. We set

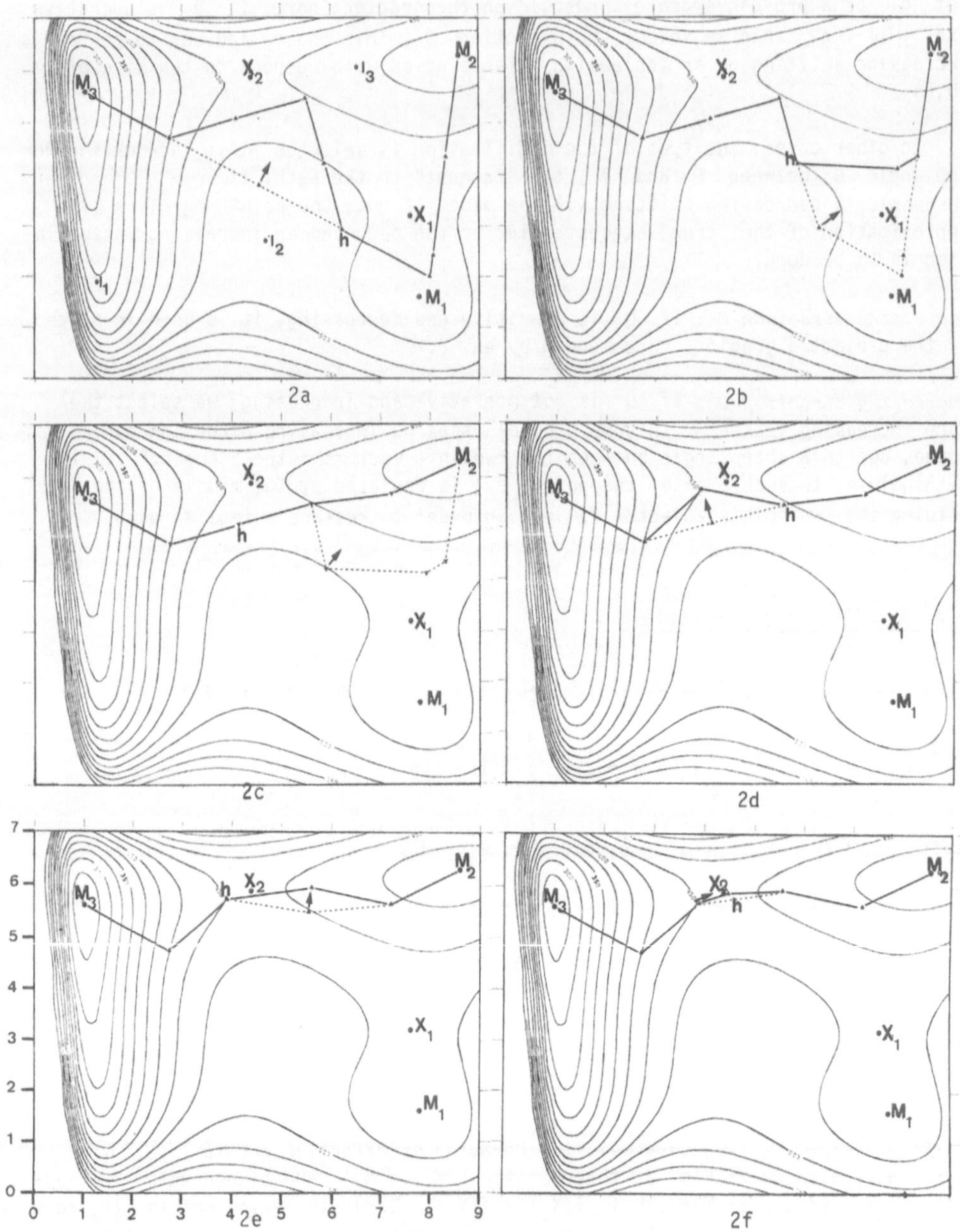

Fig. 2: ▲ successive points of the chain; h current highest point; ... previous path; ➔ current shifting direction D. (a) standard projected gradient shifting with insertion of two points; (b) identical behaviour, but X_1 is discarded when cutting-off the right chain;(c-e) projected gradient shifting and by-pass of some meanders; (f) increasing interpolation in the quadratic direction.

$$\ell_2 = 2 \; ; \quad \eta = 30° \; ; \quad g_c = 0.04 \; ; \quad \varepsilon = 0.1 \; ; \quad \ell_1 = 0.01\,\ell_2 \; ; \quad \ell_0' = 0.01\,\ell_1 \; ;$$

$$g_1 = 10\,g_c \; ; \quad r = \ell_2 \; .$$

Convergence is achieved after 26 function calls (including 8 gradient evaluation) and a good estimate of the hessian at X_2 is available.

We must keep in mind that no convergence proof of this numerical procedure has been established so far. However a lot of chemical applications in up to 16 dimensions spaces [9] always exhibit the same amazing capability to converge (in fact, no divergence has yet been observed).

References

[1] S. Glasstone, K.J. Laidler, H. Eyring: *The theory of Rate processes* (McGraw-Hill, New York 1941)

[2] D. Liotard: Int. J. Quant. Chem. Submitted for publication

[3] M.S. Berger, M.S. Berger: *Perspectives in Nonlinearity* (Benjamin, New York 1968) th. 2.25 p. 59; M.S. Berger: *Nonlinearity and Functional Analysis* (Academic New York 1977) th. 6.5.3 p. 354

[4] A. Ambrosetti, P.H. Rabinowitz: J. Funct. Anal. *14*, 349 (1973)

[5] J.N. Murrell, K.J. Laidler: Trans. Faraday Soc. *64*, 371 (1968)

[6] P. Andre: Thèse de Spécialité, Toulouse (1975); B.A. Murtagh, R.W.H. Sargent: Comp. J. *13*, 185 (1970); C.G. Broyden: Math. Comp. *21*, 368 (1967); H.Y. Huang: J. Optimization Theory Appl. 5 (1970), *6* (1970)

[7] M.J.D. Powell: In *Numerical Methods for Non-Linear Algebraic Equations*, ed. by Rabinowitz (Gordon and Breach, New York 1970) Chap. 6,7

[8] A. Komornicki, J.W. Mc Iver Jr.: J. Amer. Chem. Soc. *96*, 5798 (1974)

[9] H. Cardy, D. Liotard, A. Dargelos, E. Poquet: Nouv. J. Chim. submitted for publication; D. Liotard: Thesis, Pau (1979)

Chapter 5
Non-Physical Applications of Statistical Mechanics

Telephone Network: Statistical Mechanics and Non-Random Connecting Procedures

E. Bonomi, J.L. Lutton, and M.R. Feix

C.R.P.E., C.N.E.T., C.N.R.S.
F-45045 Orléans Cédex, France

1. Introduction

We are dealing with connecting networks as those found in telephone exchanges |6| |1|, the gross structure of which is dipicted in fig.1.

The purpose of such a system is to realize all possible connections between N inlet lines and N outlet lines ($N \sim 10^4 - 10^5$). The realization of all possible connections through the materialization of an N x N matrix would involve N^2 crossing points which is impossible to make.

The art of commutation consists in breaking this large matrix into much smaller ones linked in such a way as to shuffle the traffic and establish the maximum of wished connections. Moreover in addition to this hardware realization we must consider different strategies of routing the calls. We must provide the system with a certain intelligence (i.e. an algorithm stocked in a computer) which allows of course a saving of hardware.

The main feature of such a system is that both the control unit and the connecting network contain thousands of components giving a large number of possible combinations (microscopic states). Our purpose is to suggest that theoretical methods like those of statistical physics are especially designed to extract important macroscopic quantities such as the traffic load from thin mass of details.The problem is to build the best possible structures (which defines the microscopic phase space of the system) and the best strategies under the constraints of the macroscopic traffic data.

As already suggested by V.E. Benes |2| we study the macroscopic behaviour of the connecting system at statistical equilibrium using information theory concepts as the maximum entropy principle under constraints. In this paper, we deal only with point to point connections i.e. the N inlets and the N outlets correspon-

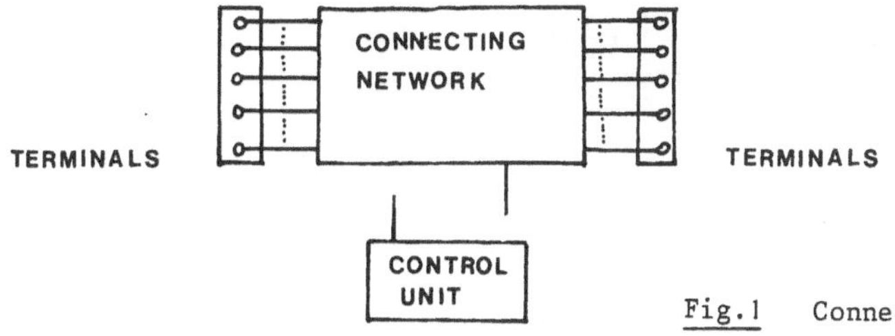

Fig.1 Connecting systems

ding to "discernible subscribers". In a first part we briefly describe the dynamical model and show in what sense and how the system reaches a thermodynamic equilibrium state in the case of a specialized class of networks (the so called Clos Networks). The second part presents some theoretical results obtained in the framework of the statistical theory and compares them with computer simulations.

In a third part we show how intelligence can be introduced in the network resulting in a saving of hardware. The most intelligent strategy involves a reconfiguration of the system.

2.1. Dynamical model

The arrivals are described by a Poisson model with all the inlet call sources working independently from each other : in such a model the probability for one idle inlet to call during time $\tau \in [\tau ; \tau + d\tau[$ is $\lambda\, d\tau$. The same model is used for the call duration with a probability $\mu d\tau$ to hang up in the interval for a busy connection. These two laws simplify considerably the analysis (and especially the computer simulation) and are much more realistic than a priori thought would be.

If a new demand of connection arrives and cannot be satisfied (this can only happen if other connections already occupy at least one link in every possible paths between the inlet and outlet in question) the demand is refused and it is supposed that the caller does not insist anymore. The probability of such a blocking is the most significant parameter of the system and must be kept sufficiently low. This probability is noted P when the traffic in stationary. It is easily shown that if P = 0 we can define a traffic demand t_D with :

$$t_D = (1 + \mu/\lambda)^{-1} \tag{1}$$

where t_D is the average percentage of inlets which attempt to have a connection. If $P \neq 0$ we have an effective traffic $t_E < t_D$ with the relation

$$t_E = \frac{t_D(1 - P)}{1 - t_D P} \tag{2}$$

t_E is now the average accupancy percentage of each inlet (outlet).

The connecting network is an arrangement of matrices allowing various combinations between itlets and outlets. Fig.2 shows the three-stage Clos network |3|. The first stage is formed of b matrices (entrance blocks) of size nxr, the total number of subscribers being N = nb. The central stage is formed of r square matrices of size bxb, the third stage being identical to the first one but with symetrical connections with respect to the central stage.

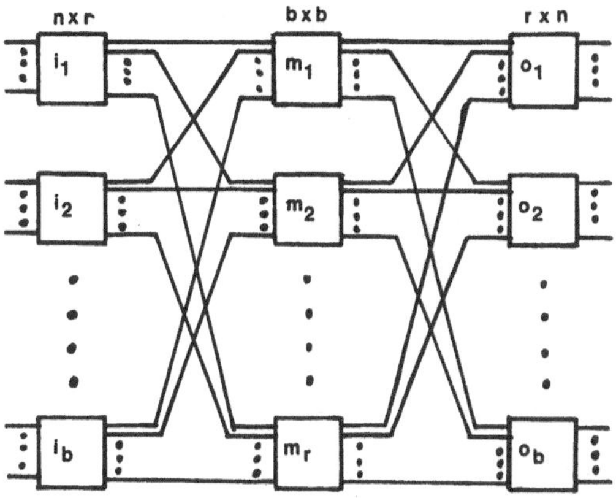

Fig.2 Three stage Clos connecting network : $r \geq n$

2.2. Thermodynamic limit and extensivity

At that point it is quite useful to represent the logical events of the three-stage Clos network by a matrix representation. We note that the connecting path is given by a triplet (u, v, w) where u, w ∈ {1, 2, ..., b} and v ∈ {1, 2, ..., r} such that there is a connection between an inlet of the entrance block i_u and an outlet of the output block O_w, via the middle block m_v. Obviously inside the two blocks i_u and O_w we do not need to specify the exact identity of the connected subscribers. This simplifies considerably the model with no loss of generality.

Hence we can represent the total system by a matrix M of size rxb. See Fig.3 with r = 3 and b = 8. The first line, first column indicates that an inlet of the entrance block I is connected to an outlet of the output block 4 through the first of the 3 central blocks labeled A, B, C. A given number appears at the most once in each raw of M and of the most n times in matrix M.

A new demand is blocked if all the lines corresponding to the free case of the entrance column already display the number of the called block.

For example a subscriber of block II can call a subscriber of block I, but not one of block 7, 8. To complete the model we must give the possible strategies to choose the central block.

	I	II	III	IV	V	VI	VII	VIII
A	4		2	5		8	7	3
B		3	5		2	1		4
C	7		3		8		5	2

Fig.3 Matrix representation of the Clos network: n = 3

1) Random Hunting. We list all free cases permitting the connection and select one of them at random.

2) Sequential Hunting. Among all these cases we select the first one available.

3) Packing rule Hunting. We select among them the one corresponding to the most heavily charged raw.

The thermodynamic limit is obtained by letting N go to the infinity together with b and keeping constant n = N/b, i.e. the size of the input and output blocks (the size of the central blocks go to infinity). Moreover we can show that in this limit the probabilities q_A, q_B, ... to get busy central matrices A, B, ... are the macroscopic quantities from which we can deduce the blocking probability $P(q_A, q_B, \ldots)$.

This result is supported by the computer simulations (fig.4) : except for small N the blocking probability is independent of b.

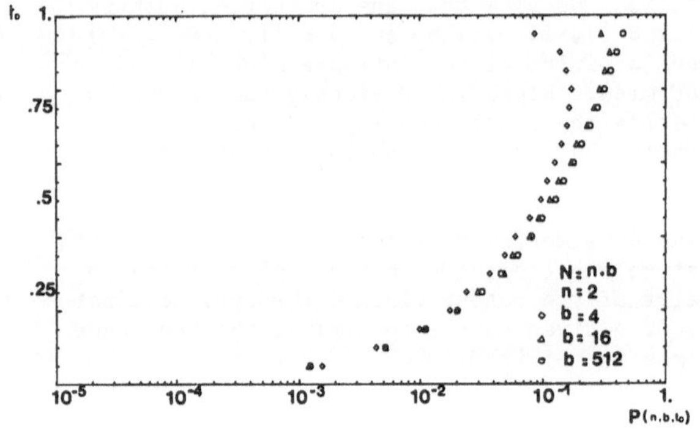

Fig.4 Concept of extensivity

3.1. Blocking probability. Random Strategy

The simplest case to treat is the random hunting strategy since now all the lines of M are equivalent and consequently have the same statistical weight. Without going into details of the calculation which could be found in |4|, we give the final results for r > n :

$$P(n,r,q) = \begin{cases} \prod_{k=0}^{r-n} \binom{n-1-k}{r-k} q^r [2-q]^{2n-r-2} & \text{if} \quad r < 2n-1 \\ 0 & \text{if} \quad r \geq 2n-1 \end{cases} \tag{3}$$

where $q(t_E)$ is the probability to find one inlet busy. Hence we take $q = t_E$ in the thermodynamic limit. Introducing this value in (3) and taking into account (2), we eliminate t_E (eventually in a numerical way) and obtain the blocking probability P as a function of the traffic demand t_D.

It could be noticed on (3) that if $r \geqslant 2n - 1$, $P = 0$. This corresponds to a well known theorem stated by Clos |3| proving that in such a case the network is strictly nonblocking. Moreover in the often considered case $r = n$, we get :

$$P(n, q) = \frac{n-1}{n} q^n \left[2 - q\right]^{n-2} . \tag{4}$$

Numerical simulations are given on fig.5. They show first the good agreement between the computer simulation results and the above formula and they exhibit the fast frop of the blocking probability when we slightly increase r over n.

The central hypothesis has been the identification of q and t_E, the average effective traffic. Systematic investigation of the problem for different values of n and r indicates that we always have :

$$t_E \lesssim q \lesssim t_D$$

with q equal to t_D for $n = r = 2$ and very close to t_E for $r > 8$.

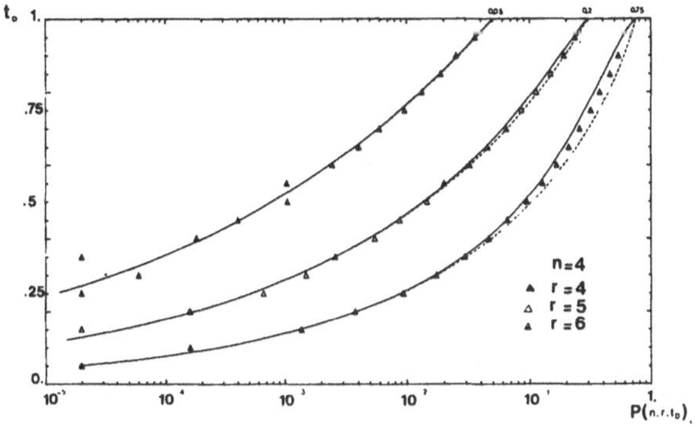

Fig.5 Blocking probabilities: —— $q = t_E$; ... $q = t_D$

3.2. Blocking probability, other strategies

In the case of other strategies we only study the case $n = r$, we obtain

$$P(n,q_1,\ldots,q_n) = \frac{n-1}{n} \sum_{j=1}^{n} q_j \left[2^{n-2} + \sum_{i=1}^{n-2} \frac{\binom{n-z}{i}}{\binom{n}{i}} (-)^i 2^{n-2-i} \sum_{n \geq s_1 > s_2 > \ldots > s_i \geq 1} q_{s_1} q_{s_2} \cdots q_{s_i}\right] \tag{5}$$

with $q = \sum_{i=1}^{n} q_i/n$ and q_i being the occupation rate of i-th line in matrix M. Unfortunately these formulas are formal since we have up to now no way to compute the q_i. Nevertheless we can use it to show one important result, namely that the random hunting strategy is the worst one : $P(n,q) \geqslant P(n, q_1, \ldots, q_n)$. On the other hand considerations on the microscopic states of matrix M indicate that a new call must be placed on the heaviest possible loaded line justifying as the best rule the packing hunting strategy. Fig.6 shows the comparison between the 3 above mentioned strategies. As predicted the random hunting is the worse while the sequential and packing rule strategies give nearly the same results with, for small traffic, a slight advantage for the packing rule, in agreement with the theoretical results. Since for not small traffic the sequential hunting strategy charges systematically, when possible, the first matrix, this last one is practically also the most loaded and the two strategies are very close to the other.

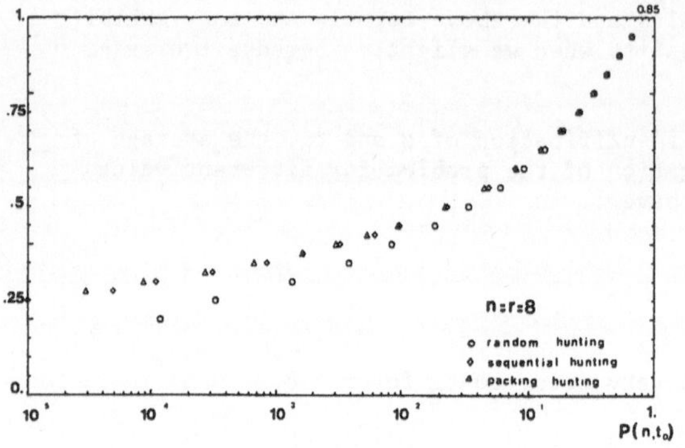

Fig.6 Comparison of the three hunting strategies

3.3. Rearrangeability

The structure of a connecting network determines in which order an assignment of inlets and outlets can be connected and simultaneously communicate. That means a given demand is blocked when the ordering of the new resulting assignment is inconsistent with the combinationial property of the system.

A rearrangeable connecting network is one the permitted states which realize every assignment of inlets and outlets, that is, one in which it is possible to rearrange existing calls so as to put in any new call.

Two important results about the three-stage Clos networks are known :

1) The Slepian-Duguid theorem, which states that such a network is rearrangeable if and only if $r \geqslant n$.

2) The theorem of Benes, which states the existence of a solution such at the most (b−1) existing calls need to be moved in order to connect an idle inlet-oulet pair.

This last result means that the upper bound of the number of removed calls increase with the size of the system, and, consequently, the amount of searching and data-processing required should become larger and more difficult. Our basic idea is to show how the concepts of thermodynamic limit and extensivity remain still useful, namely the probability of having to proceed to (b−1) reswitchings is completely negligible for large values of b.

To solve this problem we must find the probability distribution $\{P_k\}_{k=1}^{2b-3}$ where P_k is the probability to unblock the system by removing k calls in progress. If $m = m(n,r,b,t_D)$ is the average number of reswitchings, we shall maximize the entropy $S[P]$ taking m fixed :

$$S[P] = - K \sum_{k=1}^{2b-3} P_k \log P_k \tag{5a}$$

$$m = \sum_{k=1}^{2b-3} k\, P_k \, . \tag{5b}$$

The solution is in the thermodynamic limit :

$$P_k^{\infty} = \frac{1}{m_{\infty}} \exp\left[- \beta(k - 1)\right] \quad ; \quad m_{\infty} = \left[1 - \exp(- \beta)\right]^{-1} \tag{6a}$$

$$\beta(n,\ r,\ t_D) = - \log(1 - P_1^{\infty}) \, . \tag{6b}$$

The temperature β^{-1} determines all the macroscopic properties of the system, and 6b allows to measure β using the computer simulation. Fig.7 shows for $0 \leqslant t_D \leqslant 1$, the behaviour of β for the different strategies. We can see that the average number of removed calls is optimized for the random strategy which gives for r = n the following law :

$$\beta = - \log t_D \, .$$

Fig.8 shows the agreement between the theory and the computer simulation, the algorithm of rearrangability of which is described in |5| where, in the same limit, the probability to have to rearrange is given by 4-5 setting $q = t_D$.

Notice that the probability to have to rearrange is minimum in the packing rule hunting and maximum in the random hunting. On the other hand the number of calls to be moved is minimum in the random hunting strategy as can be see from Fig.7.

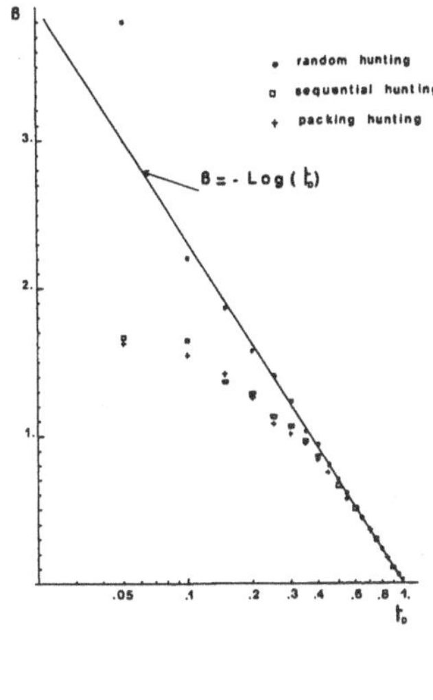

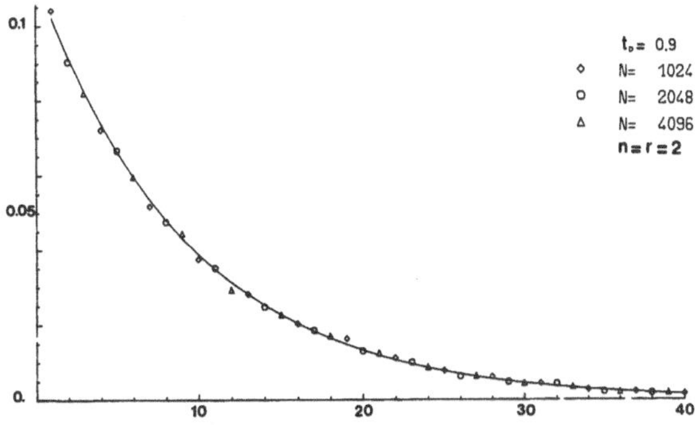

Fig.8 Probability distribution to unblock the system by removing k calls in progress (6)

Fig.7 Behaviour of β for different strategies

4. Conclusion

In this paper we have described the three-stage Clos network under different strategies and shown that it exhibits a thermodynamic behaviour through the concept of extensivity. By a systematic substitution of each central matrix by another (smaller) three-stage Clos Network we define a 5 stages network and an iteration of this process leads to the concept of multi-stage network. Again the concept of extensivity can be applied and a thermodynamic approach used to study such networks.

Both in three and multi-stage networks a sizeable saving of hardware can be obtained by the introduction of intelligence in the strategies and by the possibility of reconfigurating the calls.

To study this last point we have considered the application of the maximum entropy principle with constraints in its simpler form. We have seen this principle combined with the concept of thermodynamic limit keeps what is relevant for predicting the behaviour of the system (three-stage Clos network) and discards what is irrelevant or unlikely. The problem of reconfigurating is certainly more difficult for an high number of stages although some algorithms already exist for base 2 systems |5|.

References

|1| N. Pippenger. On cross bar Switching networks. IEEE Transaction on communications, Vol. COM-23, NO. 6, June 1975.

|2| V.E. Benes. Mathematical theory of connecting networks and telephone traffic. Academic Press. New York, 1965.

|3| C. Clos. A study of Non Blocking Switching Networks, Bell Syst. Tech. J. 32, 406-424, 1953.

|4| E. Bonomi, J.L. Lutton, M.R. Feix. A statistical mechanics approach to connecting networks : blocking probabilities, submitted for publication in NTC 80, Houston.

|5| E. Bonomi, M.R. Feix, G. Hebuterne, Théorie de l'information. Structure et modélisation d'un réseau de connection. Ann. des Télécommunications, Vol. 35, N° 1-2, Jan.-Fév. 1980.

|6| Grinsec, La commutation électronique, Edition Eyrolles, 1980.

The Thermodynamic Formalism in Population Biology

L. Dementrius
Université Paris VII, 2, Place Jussieu,
F-75005 Paris, France

Introduction

The methods of classical mechanics have dominated most areas of population biology since the work of Volterra and Lotka in the 1930's. These methods revolve around two main principles. First, the state of a system may be observationally determined; second, the future behavior of the system is completely determined by knowledge of the current state and the dynamical laws. These principles are in general valid for models in which there are few degrees of freedom and the trajectory of the system is insensitive to small changes in the initial conditions.

However, when the number of degrees of freedom of the system is large, and insensitivity to initial data no longer holds, the application of the methods of classical mechanics involves problems of both an empirical and mathematical nature. First of all in the case of large systems, determining the state of the system at some instant is a very difficult task. The most precise measurements we may make may well yield only partial information concerning the state. Second, in order to determine the trajectory of the dynamical system from the equations of motion, we must integrate a set of differential equations. This is in general a highly technical task even for systems involving few degrees of freedom. Third, in cases where small changes in the initial conditions create chaotic trajectories, the prediction of future behavior from the present state becomes practically impossible. These three issues suggest that the application of the methodology of classical mechanics to certain classes of systems can create numerous problems at various levels.

The science of statistical mechanics recognizes the problems involved in determining the precise changes that may occur in the dynamical behavior of a large system or systems with sensitive dependence on initial conditions. Statistical mechanics exploits the fact that only partial information about the state of the system is in reality possible and uses statistical methods to determine macroscopic variables that are insensitive to small changes in the initial state. This approach is illustrated by the methods which have been applied to study the macroscopic behavior of a gas. A gas consists of a very large number of molecules moving according to the laws of

*Present address: Max-Planck Institut, D-3400 Göttingen/West Germany, Postfach 968

mechanics. It is in principle impossible to determine the state of the system in the sense of determining all the positions and the velocities of the particles. According to statistical mechanics, the phase space counterpart of our gas at equilibrium at temperature T is not some point in X the set of all possible positions and velocities, but a probability measure μ on X. This equilibrium measure as proposed by Gibbs has the form

$$\mu(A) = \frac{\int_A \exp-(H/kT)dw}{\int_X \exp-(H/kT)dw}$$

where k is a universal constant, known as Boltzmann's constant and H the Hamiltonian of the system.

The fundamental principle of statistical mechanics is the

Boltzmann-Gibbs Principle: The equilibrium value of any quantity at temperature T, is the average value of that quantity in the canonical ensemble.

This principle, which enables us to compute the free energy and hence all other macroscopic variables provides a correct basis for the prediction of macroscopic behavior.

The dynamical behavior of many biological models possesses properties analogous to those which one observes in certain physical systems. For example, certain ecological models of density-dependent population growth have the property that small changes in the initial conditions can produce large effects on the dynamics of the system. These models, although deterministic in structure lead to dynamics which are indistinguishable from the sample function of a stochastic process [10]. Statistical methods abandon the attempt to determine the precise trajectory of the population density and focus instead on the mean density of the population over a certain period. This parameter along with other variables such as the growth rate and the population entropy, provide a correct basis for the prediction of macroscopic behavior [4]. Certain demographic models consist of a large number of individuals of different ages and different genotypes. These individuals are subject to different kinds of environmental interactions. In order to predict the time of death of a particular individual, we need to know the individual's age, his genetic constitution and the environmental circumstances he will confront. In addition as the number of individuals in the population increases, the characterization of the state space becomes less tractable. The methods of classical mechanics are not adequate to deal with population phenomena in which so many different variables interact. Statistical methods abandon the attempt to determine the precise trajectory of each individual and consider instead the mean number of individuals of any age in any area which can be expected to survive in a given period. Thus the mean life-expectancy of individuals in a given environment emerges as a macroscopic parameter which provides a correct basis for the prediction of the global behavior of the population.

The shift from models based on classical mechanics to models based on statistical physics seems to have been first rigorously treated in a biological context in [1]. In this paper and subsequent extensions [2,3], the methods of statistical mechanics were exploited to provide a systematic basis for the derivation of macroscopic observables in population biology. This paper reviews these methods. The main concept that dominates this review is the notion of entropy. Historically the concept entropy was introduced in classical thermodynamics by Clausius in order to provide a quantitative basis for the observation that natural processes have a particular direction. This parameter was later given a statistical interpretation by Boltzman, who defined the entropy of a system as a measure of the total number of microstates that are compatible with a given macrostate. Thus, if N denotes the number of microstates, corresponding to a given macrostate, the entropy is given by

$$S = k \log N \quad .$$

In a physical system consisting of many interacting particles, the expression which measures the degree of disorder of the system also describes the uncertainty concerning the states of the individual particles. Shannon (1948) in studying mathematical models for the transmission of information, recognized an analogy between the measure introduced by Boltzman and the degree of uncertainty associated with a stochastic process. Shannon observed that one could make precise the idea that certain probability experiments have more uncertainty and yield more information than others, and discovered that if one required that this measure of uncertainty satisfies certain conditions, then there is a unique number that characterizes the process. In 1959, Kolmogorov invented a new numerical invariant of measure preserving transformations in order to settle an important problem concerning the isomorphism of Bernoulli shifts. This invariant is called entropy since it is a generalization of the Shannon parameter. The importance of the Kolmogorov invariant in modern statistical mechanics is based primarily on its use to characterize equilibrium states in infinite systems. Dobrushin [5] has shown that the equilibrium states of certain models in statistical mechanics can be described as the equilibrium states of certain time evolutions of a Markov process. Ruelle [2], exploiting some techniques introduced by Sinai [13] has observed that the asymptotic properties of certain dynamical systems in the neighborhood of an attractor, can be characterized by an ergodic measure which satisfies a variational principle.

The work of Dobrushin, Sinai and Ruelle on infinite systems has brought statistical mechanics in close contact with the theory of stationary stochastic processes and dynamical systems. This connection naturally suggests the possibility of finding analogues of thermodynamic concepts such as free energy and entropy in other contexts besides Hamiltonian system. The program to establish such a connection in biological systems was initiated in [1]. In this work analogues of thermodynamic concepts in a demographic context were derived. It was observed that one could make precise sense out of the idea that certain populations have a more complex life-history than others. This measure of complexity was denoted entropy since it was based on the Kolmogorov invariant. This paper outlines the connection between statistical mechanics, stochastic processes and dynamical systems in order to bring out the ideas that underlie the entropy concept and other macroscopic variables that arise in the study of population phenomena.

Section 1 outlines the ideas from statistical mechanics and thermodynamics which form the background for the analysis of biological models. These methods are applied to the analysis of population models in Section 2. We derive thermodynamic analogues of population parameters and discuss the analogy between multiple equilibrium states in population models and phase transitions in physical systems.

1. Physical Systems and Configurational Thermodynamics

Consider a physical system consisting of N molecules moving in 3-dimensional space. The set of states B of the system is composed of collections

$$\bar{a} = (\bar{p}_1,\bar{x}_1) \ \dots \ (\bar{p}_N,\bar{x}_N)$$

where $\bar{p}_i$ is the momentum vector of the ith particle and $\bar{x}_i$ the velocity. The set of elements $(\bar{p}_1,\dots,\bar{p}_N)$ is called the momentum space and the set of elements $(\bar{x}_1,\dots,\bar{x}_N)$ the configuration space. The total energy H is given by

$$H = \left(\frac{1}{2m}\right)\sum \bar{p}_1^2 + W(\bar{x}_1,\dots,\bar{x}_N)$$

where $W(\bar{x}_1,\dots,\bar{x}_N)$ represents the potential of interaction between the particles. According to the fundamental postulate of statistical mechanics, the phase space counterpart of our gas in equilibrium at temperature T is not some point in X, the set of all possible positions $\bar{x}_i$ and momenta $\bar{p}_i$ but a probability measure μ in X. This probability measure according to Gibbs is

$$\mu(w) = \frac{\exp-(\beta H)dw}{\int \exp-(\beta H)dw} \qquad (1.0)$$

where w denotes the LEBESGUE measure in X associated with the coordinatization $\bar{x}_1,\dots,\bar{x}_N;\ \bar{p}_1,\dots,\bar{p}_N$ and $\beta = 1/kT$, k is a universal constant known as Boltzmann's constant. Now

$$\exp-(\beta H)dw = \prod_{i=1}^{N}\left[\exp\left(\frac{-\beta\bar{p}_i^2}{2m}\right)dp_i\right]\cdot\exp-[\beta W(\bar{x}_1\dots\bar{x}_N)d\bar{x}_1\dots d\bar{x}_N] \quad .$$

This expression factorizes into a kinetic part

$$\exp-\left(\frac{\beta\bar{p}_i^2}{2m}\right)dp_i$$

for each particle, and a configurational part

$$\exp-[\beta W(\bar{x}_1\ldots\bar{x}_N)]d\bar{x}_1\ldots d\bar{x}_N$$

which defines a configurational ensemble. Thus the Gibbs measure μ can be written as the product of two measures μ_1 and μ_2 where μ_1 is a Gaussian measure in $\mathbf{R}^{3N}$, the space of all possible momenta and μ_2 is a measure in V^N the space of all possible configurations. The partition function Z is given by

$$Z = \int \exp-(\beta H)dw = \left(\frac{2\pi m}{\beta}\right)^{3N/2} \int_{V^N} \exp-[\beta W(\bar{x}_1\ldots\bar{x}_N)]d\bar{x}_1\ldots d\bar{x}_N \quad ;$$

let us write

$$Q = \int_{V^N} \exp-[\beta W(\bar{x}_1\ldots\bar{x}_N)]d\bar{x}_1\ldots d\bar{x}_N \quad .$$

Given the partition function

$$Z = \left(\frac{2\pi m}{\beta}\right)^{3N/2} Q$$

we can in principle compute the free energy function and as a consequence derive all the other thermodynamic functions. The free energy function is

$$F = kT \log Z \tag{1.2}$$

The expressions for the thermodynamics functions, entropy S, and mean energy U, in terms of F are

$$S = -\frac{\partial F}{\partial T} \tag{1.3}$$

and

$$U = F - T\frac{\partial F}{\partial T} \quad . \tag{1.4}$$

Using (1.2) to compute the entropy S, we obtain

$$S = k \log \int \exp-(\beta H)dw + \frac{\left(\frac{1}{T}\right)\int H \exp-(\beta H)dw}{\int \exp-(\beta H)dw}$$

which yields

$$S = -k \int \rho \, \log \rho \, dw$$

where

$$\rho = \frac{\exp -(\beta H)}{\int \exp -(\beta H) dw} \quad .$$

The expression for S suggests that the Gibbs measure μ given by (1.0) can be described in terms of a variational principle. In effect, the measure μ maximizes $\int -f \log f \, dw$ subject to the condition that $\int fH \, dw$ has a fixed value U. Hence the equilibrium state in statistical mechanics can be described as the distribution which maximizes the free energy function U-ST.

Now, using (1.1), we can write the free energy F in the form

$$F = \frac{3N}{2\beta} \log\left(\frac{2\pi m}{\beta}\right) - \frac{1}{\beta} \log Q \quad . \tag{1.5}$$

The first term in this expression is due to the contribution from the kinetic energy $1/2m \sum p_i^2$, the second term corresponds to the contribution due to the potential energy.

Now, let us neglect the first term in the expression for F given by (1.5) and let define a new free energy function

$$F^* = -\frac{1}{\beta} \log Q \quad . \tag{1.6}$$

Using (1.3) and (1.4), we can compute the new entropy S^* and the new mean energy U^*. These new expressions differ from S and U by N-times known functions of T. This means that in deriving the thermodynamic functions from the energy H, we may ignore the kinetic energy and consider only the potential energy $W(\bar{x}_1,\ldots,\bar{x}_N)$. Hence, in our characterization of equilibrium states, we may replace the phase space X by V^N, the energy H by W and the measure μ by μ_2. However, the shift to the new model (V^N, W, μ_2) is not quite adequate to describe the macroscopic behavior of physical systems. In physical systems, the number of particles N is so large that is is only the limit as $N \to \infty$ that counts in answer to physical questions. A suitable model for statistical mechanics is derived by considering the case where N, the number of particles and v the volume both tend to infinity while keeping the density $d = v/N$ fixed, the so-called thermodynamic limit.

There are essentially two approaches to developing models in which the thermodynamic limit plays a role. One approach consists of computing the thermodynamic functions for finite systems of constant density and increasing volume and then taking their limits as the volume becomes infinite. Another approach consists of finding an infinite system for which these limit functions are the actual thermodynamic variables themselves. In generalizing the models of configurational thermodynamics using this approach, we replace the space V^N by some abstract measure space X which we endow with some suitable topology. The object X is the configuration space. We consider the transformation $T: X \to X$ which describes how points of the set X change in time. The energy W is replaced by some real-valued function ϕ on X. Stationary states of the system are described by measures μ which are preserved by T. Given a stationary state μ, we can define an entropy $H_\mu(T)$ which reflects the randomness of the distribution. Given an interaction $\phi: X \to \mathbb{R}$ we can define the mean energy

$$\phi_\mu = \int \phi d\mu \quad ;$$

the free energy F_μ is given by

$$F_\mu = H_\mu(T) + \int \phi d\mu \quad . \tag{1.7}$$

An equilibrium state is a state $\hat{\mu}$ such that

$$F_{\hat{\mu}} = \sup_\mu [H_\mu(T) + \int \phi d\mu] \quad . \tag{1.8}$$

In the population context we shall consider, the microscopic states are the age-specific fecundity and mortality of the individuals in the population. Each individual has a set of ancestors and descendants. We use the term genealogy to describe this set of ancestors and descendants, each such person indexed by his age. The phase space X is the set of all genealogies. The function ϕ assigns to each genealogy a number which depends on the fecundity and mortality of the individuals that describe the genealogy. The precise expression for ϕ will depend on whether or not the fecundity or mortality vary with population density. In density independent models, the birth and death rates are age-dependent but time-independent. The corresponding potential is a time-invariant two-body potential. In density-dependent models, birth and death rates are age-dependent and time-varying. The corresponding interaction is given by a many-body potential. We shall examine the existence and uniqueness of equilibrium states for these two different types of interactions. The variational principle given by (1.8) forms the mathematical basis for the analogies between thermodynamic variables and population parameters. The main analogies we derive are as follows: the temperature corresponds to the reciprocal of the generation time and the free energy corresponds to the Malthusian parameter.

2. Population Models

We describe two demographic models, a deterministic model due to Leslie [9] and its statistical representation introduced in [1]. These models only consider the female population and assume that there are enough males not to alter the birth or death rates as a function of age, of the models we are studying. In the Leslie model, the phase space is the set of all age-distributions. Equilibrium states are described by age-distributions which remain constant in time. In the statistical representation [1], the phase space is the set of all genealogies. Equilibrium states are described by shift-invariant measures on the space of all genealogies. The equilibrium distributions are characterized by a variational principle. We shall discuss the relationship between the deterministic and the statistical descriptions and also between the two notions of equilibrium. In these models, time and age are treated as discrete variables. The population is partitioned into n age-groups corresponding to the unit intervals of time. Let $x_i(t)$ denote the number of females in age-group i at time t. The proportion of females in age-group i at time t surviving to be in age-group i+1 at time t+1 is $b_i(t)$, which for $i \leq n-1$ is strictly positive. Furthermore $m_i(t)$ denotes the average number of daughters born per female to females in age-group i at time t, these daughters surviving to be in age-group 1 at time (t+1). Using these definitions, we have the following matrix equation

$$\bar{x}(t+1) = A\bar{x}(t) \tag{2.0}$$

where

$$A(t) = \begin{bmatrix} m_1(t) & m_2(t) & \cdots\cdots & m_n(t) \\ b_1(t) & 0 & \cdots\cdots & 0 \\ 0 & b_2(t) & \cdots\cdots & 0 \\ \vdots & & & \\ 0 & 0 & b_{n-1}(t) & 0 \end{bmatrix} .$$

We shall assume that for each t, $m_{n-1}(t) > 0$, $m_n(t) > 0$. A consequence of this assumption is that the matrix A(t), for each t, is irreducible and primitive. These restrictions are not the weakest conditions for primitivity to hold. The important feature of these conditions is that they are satisfied by real human population whose age structure is truncated after the last age with positive fecundity.

The dynamical model described by (2.0) is said to be at *Lotka equilibrium* if for some positive real number λ and for all $t > t_0$

$$\bar{x}(t+1) = \lambda\bar{x}(t) \quad . \tag{2.1}$$

Thus, at Lotka equilibrium, the relative number of individuals in each age-group remains constant in time. The case of a constant fecundity m_i and a constant mortality b_i is of particular interest. When this condition holds, an equilibrium state exists and is unique. The equilibrium state $\bar{x}$ is the right eigenvector of the constant matrix A. We have

$$A\bar{x} = \lambda\bar{x} \quad . \tag{2.2}$$

The eigenvalue λ is the spectral radius of A. The parameter $\log\lambda$, which we denote r, represents the rate of increase of the total population size. This parameter is called the Malthusian parameter. Also, the population converges towards the equilibrium state $\bar{x}$, convergence here is interpreted to mean that

$$\lim_{k\to\infty} \frac{A^k}{\lambda^k} \bar{y} = c\bar{x}$$

where $y = (y_i)$, $y_i \geq 0$ is any initial age-distribution.

The statistical representation of the dynamical model described by (2.0) is obtained by first associating a graph G with the matrix A. This graph consists of n nodes. We join node i to j if $a_{ji} > 0$ where $A = (a_{ij})$ represents the i,j entry of the matrix. We have the graph

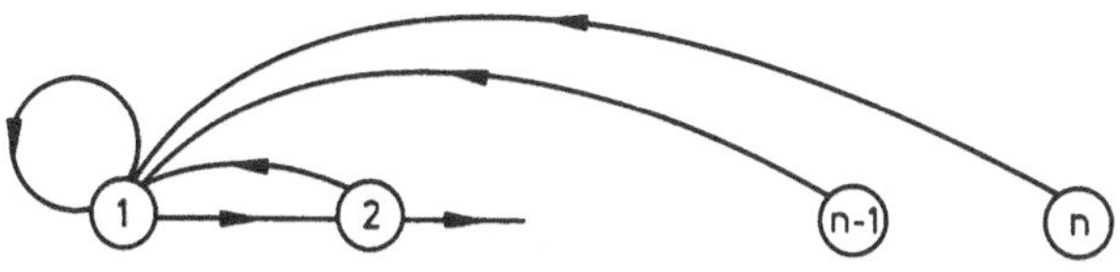

where

denotes the aging process and

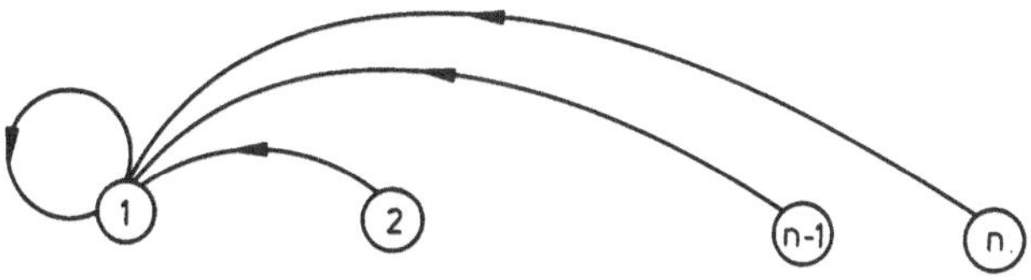

denotes the reproductive process.

Let Y denote the set of all doubly infinite paths of the graph G. The set Y is the phase space or configuration space of our system. An element $x \in Y$ is a doubly infinite sequence

$$x = (\ldots\ldots x_{-1}x_0x_1\ldots\ldots)$$

where the x_i can assume values between 1 and n. We shall call each element $x \in Y$ a genealogy.

A suitable topology in Y is induced by the metric

$$d(x_n,y_n) = \sum_{n=-\infty}^{\infty} \frac{|x_n-y_n|}{2^{|n|}}$$

which gives a measure of distance between genealogies. Let

$$T : (x_k) \rightarrow (x_k')$$

where

$$x_k' = x_{k+1} \quad .$$

The equilibrium measures we shall consider will be probability measures that are invariant under the transformation T.

To describe the class of potentials for our population model, we consider a finite subset J of Z, the integers. Let Y_J denote the set of mappings of the set J into the set $S = (1,2...n)$. This set Y_J can be described as

$$Y_J = \prod_{i \in J} S$$

which is the direct product of $|J|$ copies of the set S, where $|J|$ denotes the cardinality of J. We will denote the elements of the space Y_J by the symbol x_J. Then we have the representation

$$x_J = (x_{i_1} \ldots x_{i_{|J|}}) \quad \text{with} \quad x_{i_k} \in S$$

and

$$J = (i_1, i_2 \ldots i_{|J|}) \quad .$$

The elements in the set Y_J can be considered as restricted configurations of the set

$$X = \prod_{i \in Z} S \quad .$$

Now the phase space of our population model is the set Y and not the set X. We take this into account by choosing for each $J \subset Z$, a subset $\tilde{Y}_J$ of the space Y_J, the elements in the set $\tilde{Y}_J$ being the restricted configurations of the set Y_J.

An interaction on the phase space is defined as follows. This is any real valued continuous function on the space

$$\bigcup_{J \subset Z} \tilde{Y}_J$$

of all configuration spaces over the finite subsets J which satisfies the following conditions

(a) $\phi(\tilde{Y}_\emptyset) = 0$, where $\emptyset$ denotes the empty set in Z

(b) For all $k \in Z$, the quantity

$$\|\phi\|_k = \sum_{J,k\in J} \frac{1}{|J|} \sup_{x_J \in \tilde{Y}_J} |\phi(x_T)| < +\infty$$

This last condition means that the function ϕ should decrease fast enough with the diameter of the set J.

We introduce some terminology which will facilitate our discussion of interactions that arise in population models. An interaction ϕ is said to be an *n-body potential* if

$$\phi(x_T) = 0 \quad , \quad \text{for all } x_J \in Y_T \text{ with } |J| > n \quad .$$

An interaction ϕ is said to be of *finite range* if there exists a real number k with

$$\phi(x_T) = 0 \qquad \text{for diam } J = \max|i-j| > k \quad , \quad i,j \in J \quad .$$

Translation invariant interactions are functions ϕ which satisfy the relation

$$\phi(x_J) = \phi(Tx_J)$$

where T denotes the translation operator.

The population interactions we consider distinguish between density-independent and density-dependent models.

(A) *Density-Independent Models:*

These models are described by a time-independent fecundity and mortality and an age-distribution which at time t, is uniquely determined by the age-distribution at time (t-1). The time independence implies that the interaction ϕ is a translation invariant two-body potential, that is

$$\phi(x) = \phi_0(x_0,x_1) + \phi_1(x_0,x_2) + \dots \quad .$$

The second condition implies that ϕ is finite range, that is

$$\phi_k(x_0,x_k) = 0 \quad \text{for} \quad k > 1 \quad .$$

In general, for density-independent interactions, without time lag, we have the potential

$$\phi(x) = \phi_0(x_0, x_1) \quad .$$

This class of potentials, as we will note, gives rise to unique equilibrium states.

(B) *Density-Dependent Models:*

These models are described by an age-specific fecundity and mortality which are time-dependent. On account of the time-dependence, the interaction ϕ is no longer translation invariant. The interaction ϕ is given by a many-body potential which assumes the form

$$\phi(x) = \phi_0(x_0) + \phi_1(x_0, x_1) + \phi_2(x_0, x_1, x_2) + \dots \quad .$$

This class of interactions will in general give rise to multiple equilibrium states.

Our analysis of equilibrium states is based on the following circle of ideas. Good references to these ideas are the works of Ruelle [12] and Mayer [11]. Let

$$X^+ = \prod_0^\infty S$$

where

$$S = (1,2,\dots,n)$$

and let

$$Y^+ = \left\{x \in X^+ : a_{x_i, x_{i+1}} > 0 \quad \text{for each } i\right\}$$

where $A = (a_{ij}) \geq 0$ denotes the population matrix.

We give S, the discrete topology and X^+, the product topology. The transformation $T : (x_k) \to (x_k')$ is a finite-to-one continuous map of Y^+ onto itself.

Let $C(Y^+)$ denote the Banach space of all real-valued continuous functions on Y^+. Given an interaction $\phi \in C(Y^+)$, we define the operator

$$L_\phi : f(x) \to \sum_{y \in T^{-1}(x)} \exp[\phi(y) f(y)] \quad .$$

Let I^+ denote the set of all Borel probability measures on Y^+. The function ϕ is said to satisfy the Perron-Fröbenius condition if there exist $\lambda > 0$, $h \in C(Y^+)$ with $h > 0$ and $g \in I^+$ for which

(i) $L_\phi h = \lambda h \quad ,$

(ii) $L_\phi^* g = \lambda g \quad \text{with } g(h) = 1 \quad ,$

(iii) $\left\| \frac{L_\phi^n(f)}{\lambda^n} - g(f)h \right\| \to 0$, for every $f \in C(Y^+)$.

Now if ϕ satisfies the Perron-Fröbenius condition, then the measure

$$\hat{\mu}(f) = g(hf)$$

is a translation invariant, and

$$H_{\hat{\mu}}(T) + \int \phi d\hat{\mu} = \sup_{\mu}[H_{\mu}(T) + \int \phi d\mu] \quad .$$

This means that $\hat{\mu}$ is an equilibrium state and is indeed the unique equilibrium state for ϕ. Moreover, this equilibrium state is a Gibbs state, that is, there exist constants $c_1 > 0$, $c_2 > 0$ and a constant P, such that

$$c_1 \leq \frac{\mu\{[x_0\ x_1 \ldots x_{m-1}]\}}{\exp[-Pm + \phi(T^k(x))]} \leq c_2$$

for every $x \in Y^+$ and every $m > 0$.

When the Perron-Fröbenius condition fails, which is the case for a certain class of many-body potentials, multiple equilibrium states may exist. These facts will be the central results we exploit in the discussion that follows.

Density-Independent Models: Unique Equilibrium States

The potential function for these models are translation-invariant, finite range two-body potentials of the form

$$\phi(x) = \phi_0(x_0, x_1) \quad .$$

When the dynamical description is given by the equation $\bar{x}(t+1) = A\bar{x}(t)$ where A denotes the population matrix, the appropriate interaction is

$$\phi(x) = \log a_{x_0 x_1} \quad .$$

The operator

$$L_\phi : C(Y^+) \to C(Y^+)$$

reduces to an operator

$$L_\phi : \mathbb{R}_n^+ \to \mathbb{R}_n^+$$

and is in effect the transition matrix

$$A = \begin{bmatrix} m_1 & m_2 & \cdots & m_n \\ b_1 & 0 & \cdots & 0 \\ 0 & b_2 & \cdots & 0 \\ \cdot & & & \\ \cdot & & & \\ \cdot & & & \\ 0 & 0 & b_{n-1} & 0 \end{bmatrix} .$$

This matrix is irreducible and primitive. This means that the Perron-Fröbenius condition is satisfied, hence there exists a unique measure $\hat{\mu}$ such that

$$F_{\hat{\mu}} = \sup_{\mu}[H_{\mu}(T)+\int \phi d\mu] \quad .$$

The equilibrium measure $\hat{\mu}$ is a Markov measure and is a probability measure on the space of genealogies described by the phase space Y. We can compute $H_{\hat{\mu}}(T)$ and $\int \phi d\hat{\mu}$ for this model. We first give expressions for the eigenvectors $\bar{u}$, $\bar{v}$ of the matrix A. Let

$$l_j = \begin{cases} 1 & j = 1 \\ \prod_{r=1}^{j-1} b_r & j > 1 \end{cases} .$$

The vectors $\bar{u} = (u_i)$ and $\bar{v} = (v_i)$ defined by

$$A\bar{u} = \bar{u}$$
$$\bar{v}A = \bar{v}$$
$$(\bar{u},\bar{v}) = 1$$

are given by

$$u_i = \frac{l_i}{\lambda^i}$$

and

$$v_i = \frac{\sum_{j=i}^{n} m_j u_j}{T u_i}$$

where

$$T = \sum_{j=1}^{n} j m_j u_j \quad .$$

Let

$$p_{ij} = \frac{a_{ij}u_i}{u_j} \quad .$$

The matrix $P = (p_{ij})$ is a stochastic matrix and is in effect the probability matrix that describes the Markov measure $\hat{\mu}$. We have for the matrix P

$$P = \begin{bmatrix} p_1 & p_2 & \cdots & p_n \\ 1 & 0 & \cdots & 0 \\ 0 & 1 & \cdots & 0 \\ \cdot & & & \\ \cdot & & & \\ 0 & 0 & 1 & 0 \end{bmatrix}$$

where

$$p_j = \frac{l_j m_j}{j} \quad .$$

The expression p_j is the probability distribution for the age of reproducing individuals. The entropy $H_{\hat{\mu}}(T)$ is precisely the entropy of the Markov process characterized by the probability matrix P. Let $q = (q_i)$ denote the stationary distribution of P. Then the entropy is given by

$$H = - \sum_i \sum_j q_i p_{ij} \log p_{ij}$$

which yields

$$H = - \frac{\sum p_j \log p_j}{\sum jp_j} \quad . \tag{2.3}$$

The expression H measures the variability of the contribution of the different age-classes to the equilibrium age-distribution.

The expression in the denominator of H, which we denote

$$T = \sum jp_j$$

is the generation time.

The mean energy and the free energy can also be computed. We have

$$F_{\hat{\mu}} = \log\lambda$$

and

$$\bar{\phi}_{\hat{\mu}} = \int \phi d\hat{\mu} = \frac{\sum p_j \log l_j m_j}{\sum jp_j} \quad . \tag{2.4}$$

The expression for ϕ which we denote $\bar{\phi}$ measures the mean of the contribution of the different age-classes to the growth rate.

These expressions are related, as should be obvious from our definition of equilibrium state. We have

$$\log\lambda = H + \bar{\phi} \tag{2.5}$$

let

$$H^* = -\sum p_j \log p_j \quad .$$

The expression (2.5) becomes

$$\log\lambda = \frac{H^*}{T} + \bar{\phi} \quad . \tag{2.6}$$

The thermodynamic analogues of our population parameters emerge when we compare (2.6) with the classical thermodynamic relation

$$F = -S\bar{T} + U \quad . \tag{2.7}$$

In this expression, F is the free energy, $\bar{T}$ the temperature, S the entropy and U the mean energy. We observe that the temperature corresponds to the reciproçal of the generation time and the free energy corresponds to the growth rate.

The intuition that underlies these analogies can be understood by distinguishing between two kinds of macroscopic variables in population dynamics. The first class of variables, analogous to the intensive variables in thermodynamic includes quantities which are proportional to the size of the system occupied. Typically these are the number of individuals in the population, and the volume or area these individuals occupy.

The second class of variables analogous to the extensive quantities, consists of variables having at every point a well defined value which remains constant when the volume or area occupied is increased. Typical examples are the population density, the growth rate and the generation time. Now, the biologist measures the growth rate say, of a colony of organisms by sampling a fixed volume at two instants of time. This empirical growth rate will be less dependent on the size of the volume sampled provided the volume is sufficiently large. This suggests that the theoretical quantity to compare with experimental data is the value of the growth rate when the population size and the volume occupied by the colony of organisms both tend to infinity, the density remaining constant. This justification of the limiting process in population models, corresponds to one of the reasons given for considering the asymptotic properties of the parameters in thermodynamic theory. The formal equivalence of these two limiting processes underlies the significance of the analogues we derive between population variables and thermodynamics theory.

Density-Dependent Models: Multiple Equilibrium States

In general a substance can exist in different physical states such as gas, liquid or solid. These states are sometimes referred to as the gaseous phase, the liquid phase or the solid phase, respectively. In many physical situations at equilibrium, slight changes of some macroscopic variables can induce drastic changes in phase, so-called phase transitions. For example, in the case of a typical pure fluid such as argon, the fluid exists as a gas at fairly low densities. If the pressure is increased at a fixed temperature, the density of the gas increases but it remains homogeneous. However, when the pressure reaches the vapor pressure, some of the gas begins to condense into a denser liquid phase. At this pressure, gas and liquid coexist at thermal equilibrium, the relative amounts of the two phases being arbitrary, almost all the fluid may be present as liquid or as vapor or the two phases may occupy comparable volumes.

The situation involving changes in phase is summarized by the Gibbs phase rule. Let us define the number of degrees of freedom of the system as the number of variable factors, temperature, pressure and density which must be arbitrarily fixed in order that the condition of the system be properly defined. Then a gas or vapor has two degrees of freedom. The system liquid-vapor has one degree of freedom, since, if a system consists of a liquid in contact with its own vapor, then the condition of the system becomes perfectly defined by giving one of the variables a fixed value. Liquid and vapor can coexist under a given pressure only at a given temperature. The system solid-liquid-vapor has no degree of freedom since if these three phases coexist, then none of the independent variables (temperature, pressure, density) can be altered without destroying the nature of the system.

The phenomenon we have described relates the number of degrees of freedom of a system consisting of a single component with the number of coexisting phases.

The Gibbs phase rule generalizes this to an arbitrary number of components. Let F denote the number of degrees of freedom, C the number of components, P the number of phases. Then

$$F = C - P + 2 \quad .$$

The notion of coexistence of phases has analogues in population biology. Populations show immense variability in their life-history. Certain individuals, called semelparous organisms, breed only once in their life time. Other organisms, the iteroparous types, breed repeatedly. Within a single population, it is possible to find a mixture of both types, the composition of this mixture being quite arbitrary. The types can be interpreted as phases and shifts from one type to another as certain population parameters change can be considered as phase transitions. A simple example of this phenomenon is given by a certain species of locust. Members of this species can exist in two forms called Solitaria and Gregaria, these two forms being characterized by different life-histories. The type Solitaria has a slow developmental rate, is highly iteroparous and grows to a large size. The type Gregaria, which are migratory swarmers, hatches sooner, has a shorter life, matures rapidly and is less iteroparous.

The mechanism for transition from Solitaria to Gregaria is due primarily to crowding; locusts remain in phase Solitaria until their density reaches a certain critical value. The physical nature of the transition is a change from an assembly of finite sized clusters, the solitary locusts, to a state in which effectively only one macroscopically large cluster is present, namely the swarmers.

The analogue of the Gibbs phase rule in an ecological setting is less clear. We conjecture that it corresponds to the relationship between the number of species that can coexist for a given number of resources. We shall say no more about this aspect as the analogy is not well understood. We shall focus on the phase transition phenomena and consider the class of interactions that can yield coexistence of different life-history types.

We now examine a class of potentials that give rise to multiple equilibrium states in populations. As is now known, phase transitions in one-dimensional models of statistical mechanics can arise either in the case of pair interaction potentials that violate a strong short range conditions, Dyson [6]; or in the case of many-body interactions of indefinitely high order, Felderhof and Fisher [7]. Now translation-invariant pair potentials correspond in the population context to interactions that describe density-independent models. In this class of models in which time lag effects may occur the population matrices are constant, and the age-distribution at time (t) depends on the age-distribution at time (t-1), (t-2), ... and so on.

The many-body potentials of indefinitely high order correspond in the population context to models in which the age-specific fecundity and mortality are time-dependent, this time-dependence arising possibly from density-dependent effects. We shall consider this second class of potentials in our analysis of phase transitions. In our detailed choice of the interactions we will be guided by our insight into the origin of life-history transitions among the types Solitaria and Gregaria. The mathematical analysis of our model is based on ideas due to Felderhof and Fisher [7] and some recent results due to Hofbauer [8].

We consider a population divided into two age-groups. The interaction is given by the graph

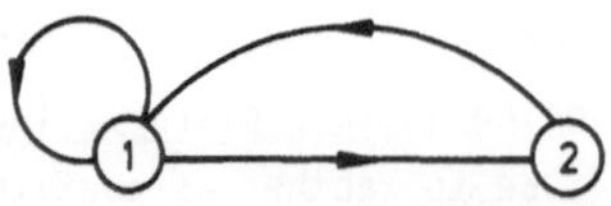

We will assume that the fecundity is constant and the mortality time-dependent. The potential ϕ on the space of genealogies Y is given by a many-body interaction which has the form

$$\phi(x) = \phi_0(x_0) + \phi_1(x_0x_1) + \phi_2(x_0x_1x_2) + \dots$$

where the ϕ_i are defined as follows

$$\phi_0(x_0) = \begin{cases} a_0 & , \quad x_0 = 2 \\ 0 & , \quad x_0 = 1 \end{cases}$$

$$\phi_1(x_0x_1) = \begin{cases} a_i & , \quad x_0x_1 = 12 \\ 0 & , \quad x_0x_1 \neq 12 \end{cases}$$

$$\phi_k(x_0x_1\dots x_k) = \begin{cases} a_k & , \quad x_0x_1\dots x_k = 11 \dots 12 \\ 0 & , \quad x_0x_1\dots x_k \neq 11 \dots 12 \end{cases} .$$

We will assume that the $\{a_k\}$ are negative numbers with $\lim a_k = 0$. This condition on the sequence a_k, in the context of statistical mechanics, ensures that the free energy in the thermodynamic limit satisfies the appropriate convexity and stability properties. In biological terms, the assumption on a_k means that the individual survivorship increases as the population evolves. This condition on survivorship, in addition to the constant fecundity, ensures the existence of an asymptotic growth rate. Let

$$S_k = a_0 + a_1 + \dots + a_k \quad .$$

We shall now impose various conditions on the sequence S_k to obtain unique and multiple equilibrium states. The proofs of these results depend essentially on the spectral properties of the Perron-Fröbenius operator L_ϕ.

The operator L_ϕ for the interaction we have described is given by the infinite matrix

$$\begin{bmatrix} 0 & e^{a_0} & e^{a_0} & e^{a_0} & \dots \\ e^{a_1} & 0 & 0 & 0 & \dots \\ 0 & e^{a_2} & 0 & 0 & \dots \\ 0 & 0 & e^{a_3} & 0 & \dots \\ \cdot & \cdot & \cdot & \cdot & \dots \end{bmatrix}$$

The associated graph of this matrix is:

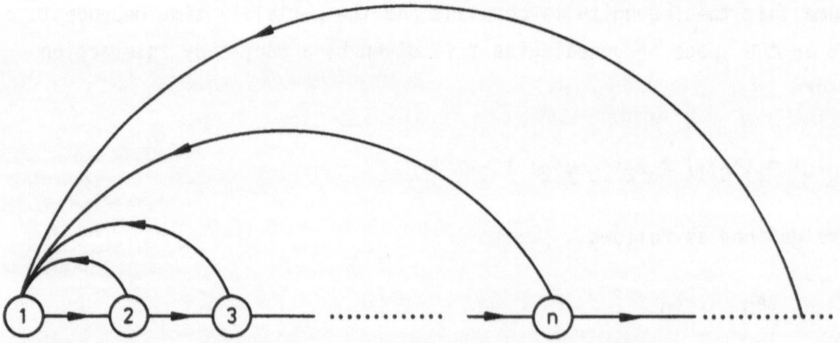

This matrix can be considered as describing a population with an infinite number of age-classes, with each age-class having constant fecundity e^{a_0} and the survivorship from age class (i) to (i+1) denoted e^{a_i}. This description will be useful in interpreting biologically the analytical conditions we will impose on the sequence S_k.

(A) Suppose $\sum \exp(S_k)>1$. This inequality implies that the operator L_ϕ satisfies the Perron-Fröbenius condition. Hence a unique equilibrium state exists. The above inequality means that the population has a positive growth rate.

We now consider two cases:

(i) $\sum a_k$ converges. The equilibrium state μ in this cases is ergodic and is a Gibbs state. The mathematical condition on the sequence (a_k) corresponds to a population in which all individuals have a finite life-expectancy.

(ii) $\sum a_k$ diverges. In this case ergodicity of the equilibrium state breaks down and the measure μ is not Gibbsian. The analytical condition on (a_k) corresponds to a population in which some of the individuals have an infinite life-expectancy. This situation will destroy the ergodicity of the genealogical distribution and create two kinds of genealogies, one containing individuals with finite life expectancy, the other consisting of individuals with infinite life expectancy.

(B) Suppose $\sum \exp(S_k) = 1$. The operator L_ϕ in this case does not satisfy the Perron-Fröbenius condition. This analytic condition means that the asymptotic growth rate of the population is zero.

We shall also examine two cases:

(i) $\sum(k+1) \exp(S_k) < \infty$. In this case the potential ϕ admits two equilibrium states μ_1 and μ_2 both ergodic. We have for both μ_1 and μ_2

$$H_\mu(T) + \int \phi d\mu = 0 \quad .$$

The restriction on S_k corresponds to a population with finite mean generation time. The measure μ_1 which is the Dirac measure concentrated at age group (1), corresponds to a semelparous population, that is a population in which reproduction is concentrated at a single instant. The equilibrium state μ_2 describes an iteroparous population, that is one in which reproduction occurs at several instances in the individual life cycle.

(ii) $\sum(k+1)\exp(S_k) = \infty$. This condition gives rise to a unique equilibrium state which is the Dirac measure. The restriction on the sequence (S_k) corresponds to a population with infinite generation time. The Dirac measure means that the population has zero entropy. In this case we have a population characterized by zero growth rate, infinite generation time and zero entropy.

The conclusions we have drawn from the two cases analyzed in (B) are compatible with the analogy we have derived between generation time and the reciprocal of the temperature. For example, perfect crystals have a very orderly structure and at very low temperatures the lattice vibration will all be in their lowest state which corresponds to zero energy. Thus a crystal will have a very low entropy at temperatures approaching the absolute zero.

Organisms in the dormant phase are the biological analogues of perfect crystals. During the dormant phase there is no metabolic activity. Organisms in this phase may be considered as having infinite generation time, zero growth rate and zero entropy.

References

1. L. Demetrius: Demographic parameters and natural selection. Proceedings National Academy of Sciences *71*, 4645-4647 (1974)
2. L. Demetrius: Adaptive value, entropy and survivorship curves. Nature *275*, 231-232 (1978)
3. L. Demetrius: La valeur adaptative et la sélection naturelle. Comptes Rendus de l'Académie des Sciences *290*, 1491-1494 (1980)
4. L. Demetrius: Macroscopic parameters in complex systems (to appear 1980)
5. R. Dobrushin: The problem of uniqueness of a Gibbsian random field and the problem of phase transitions Functional Analysis and Applications *2*, 302-312 (1968)
6. F. Dyson: Existence of a phase transition in a one-dimensional Ising ferromagnet. Communications in Mathematical Physics *12*, 91-107 (1969)
7. B. Felderhof, M. Fisher: Phase transitions in one-dimensional cluster interaction fluids. Annals of Physics *58*, 176-300 (1970)
8. F. Hofbauer: Examples for the non-uniqueness of the equilibrium state. Transactions of the American Mathematical Society *228*, 223-241 (1977)
9. P.H. Leslie: On the use of matrices in population mathematics. Biometrika *33*, 183-212 (1945)
10. R. May: Simple mathematical models with very complicated dynamics. Nature *261*, 459-466 (1976)
11. D. Mayer: The Ruelle-Araki Transfer Operator in Statistical Mechanics, in Lecture Notes in Physics, Vol.123 (Springer, Berlin, Heidelberg, New York 1980)
12. D. Ruelle: Thermodynamic formalism (Addison-Wesley 1978)
13. Y. Sinai: Markov analysis and C-diffeomorphisms. Functional Analysis and Applications *2*, 64-89 (1968)

Asymptotic Inference for Markov Random Fields on $\mathbb{Z}^d$

J. Demongeot
Université de Grenoble I, Laboratoire d'Informatique et de Mathématiques Appliquées, BP : 53 X, F-38041 Grenoble Cédex, France

1. Introduction

The purpose of this paper is to generalize results of D.K. PICKARD [1,2,3] on asymptotic inference for the Ising model. Using properties of Gibbs measures on $\mathbb{Z}^d$, we give the canonical exponential structure in which we can study the estimation problem, in the case of vertical sampling. This exponential structure depends on the range of the interaction potential of the underlying Markov random field; we construct, therefore, a test to measure this range: this test is based on the markovian character of the associated Gibbs measure; next we present the properties and the asymptotic laws of the estimators and we compare horizontal to vertical sampling [4]. Certain properties depend on the criticality of the interaction potential [5,6]: consequently, we construct a test of criticality for the interaction potential. Finally, we set the problem of estimation in a random Ising model.

This inference can be used in the scientific areas, which need random fields for the modelling of their phenomena: for example, in medicine, epidemiology at equilibrium (endemiology) and carcinogenesis [7]; in molecular biology, denaturation-renaturation of nucleic acids and folding of proteins [8]; in physics, spontaneous magnetization and spin glasses [9]. In these fields of application, the observation of such random fields can be obtained from experiments or from simulations (it is possible to simulate Gibbs measures as asymptotic measures of markovian processes). The inference allows the estimation of the potential of the field.

2. Definitions

2.1 Random Field

We call *random field on* $\mathbb{Z}^d$ a family $X=\{X_z\}_{z\in\mathbb{Z}^d}$ of random variables, indexed by $\mathbb{Z}^d$. We suppose here that we have:

$\{0,1\}^{\mathbb{Z}^d}$

is the canonical set of the trajectories of X and we identify this set with $P(\mathbb{Z}^d)$, the set of all subsets of $\mathbb{Z}^d$. $P(\mathbb{Z}^d)$ will be provided with its borelian σ-algebra $B(\mathbb{Z}^d)$ generated by the discrete topology, i.e., by the cylinders:

$$[A,B] = \{C \subset \mathbb{Z}^d \,; C \cap B = A\} \quad , \quad \text{where } B \subset \mathbb{Z}^d, \ |B| < \infty \text{ and } A \subset B \quad .$$

The subset B is called the basis of the cylinder [A,B]. We denote by $F(\mathbb{Z}^d)$ the set of subsets B of $\mathbb{Z}^d$, whose cardinality $|B|$ is finite.

2.2 Gibbs Measure [10]

The measure of X on its canonical space $(P(\mathbb{Z}^d), B(\mathbb{Z}^d))$ is called *Gibbs measure with interaction potential* V, if we have:

$$\forall\, B \in F(\mathbb{Z}^d),\ A \subset B,\ \mu([A,B] \mid B(B^c))(G) = F_B(G) \exp\left(\sum_{\substack{Y \in F(A \cup G) \\ Y \cap A \neq \emptyset}} V(Y) \right) \quad ,$$

where this equality holds for μ-almost every G in $P(\mathbb{Z}^d)$ — and where we have:

$$\mu([A,B] \mid B(B^c))$$

denotes the probability of the cylinder [A,B], conditional to the σ-algebra $B(B^c)$ generated by the cylinders, whose basis is in B^c.

Note that F_B is the random variable $B(B^c)$-measurable defined by:

$$\forall\, G \subset \mathbb{Z}^d,\ F_B(G) = \left(\sum_{B \supset A \neq \emptyset} \exp\left(\sum_{\substack{Y \in F(A \cup G) \\ Y \cap A \neq \emptyset}} V(Y) \right) + 1 \right)^{-1} \quad .$$

V is a real function defined on $F(\mathbb{Z}^d)$, which is $B(\mathbb{Z}^d)$-measurable and verifies the following condition, if $\mathbb{Z}^d = \{z_i\}_{i \in \mathbb{N}}$: the serie

$$\{a_n\}_{n \in \mathbb{N}} = \left\{ \sum_{z_n \in A \subset \{z_1, \ldots, z_n\}} V(A) \right\}_{n \in \mathbb{N}}$$

converges for every indexation of $\mathbb{Z}^d$, and we assume that $V(\emptyset)$ is equal to 0.

V is called *uniform*, if we have:

$$\|V\| = \sup_{z \in \mathbb{Z}^d} \sum_{z \in B \in F(\mathbb{Z}^d)} \sum_{A \subset B} |V(A)| < \infty \quad .$$

In practice, there exists a subset M in $\mathbb{Z}^d$, such that V vanishes except on the translated subsets of M. Then we say that the *range* of V is M. We can also define the *potential* U associated with V as follows:

$$\forall\, D \subset \mathbb{Z}^d,\ U(D) = \sum_{E \in F(D)} V(E) \quad .$$

Note that we have reciprocally:

$$\forall D \in F(\mathbb{Z}^d),\ V(D) = \sum_{C \subset D} (-1)^{|D \setminus C|} U(C) \ .$$

2.3 Markov Random Field [5,11]

A random field X is called *M-Markov random field*, if its canonical measure μ verifies:

(i) $\forall B \in F(\mathbb{Z}^d)$, $A \subset B$, $\mu([A,B]|B(B^c))$ is continuous and strictly positive;

(ii) if, for every subset Y of $\mathbb{Z}^d$, we denote by ∂Y the frontier of Y defined by: $\partial Y = \left(\bigcup_{z \in M} (Y+z) \setminus Y \right)$, we have:

$$\forall z \in \mathbb{Z}^d, \forall B \subset \mathbb{Z}^d,\ \partial\{z\} \subset B \Rightarrow \mu([\{z\},\{z\}]|B(B)) \overset{\mu\text{-a.s.}}{=} \mu([\{z\},\{z\}]|B(\partial\{z\})) \ .$$

In addition, if the conditional probabilities above are invariant for every translation of z and B, the field is called *homogeneous*.

2.4 Supermodular Potential

U is called *strongly supermodular* if, for any finite family $\{A_i\}_{i=1,\dots,n}$ in $P(\mathbb{Z}^d)$, Poincaré's inequality holds:

$$U\left(\bigcup_{i=1}^{n} A_i\right) \geq \sum_{i=1}^{n} U(A_i) - \sum_{i<j} U(A_i \cap A_j) - \dots - (-1)^k \sum_{i_1 < \dots < i_k} U(A_{i_1} \cap \dots \cap A_{i_k})$$
$$- (-1)^n U\left(\bigcap_{i=1}^{n} A_i\right) \ .$$

It is easy to see that U is strongly supermodular if, and only if, the associated interaction potential V is *attractive*, i.e., if we have:

$$\forall A \in F(\mathbb{Z}^d),\ |A| \geq 2 \Rightarrow V(A) \geq 0 \ ;$$

U is called *supermodular* if Poincaré's inequality holds only for $n = 2$.

2.5 Equilibrium State

A probability on $(P(\mathbb{Z}^d), B(\mathbb{Z}^d))$ is called *equilibrium state* for the potential U, if we have, for μ-almost every G in $P(\mathbb{Z}^d)$:

$$\forall B \in F(\mathbb{Z}^d),\ A \subset B,\ \mu([A,B]|B(B^c))(G) = \exp(U(A \cup (G \setminus B)) - U(G \setminus B))\,\mu([\emptyset,B]|B(B^c))(G) \ .$$

3. Some Results about Gibbs Measures

The following result characterizes Gibbs measures and proceeds from a property of its conditional probabilities.

Proposition 3.1

μ is a Gibbs measure if, and only if, we have:

(i) $\forall\, B \in F(\mathbb{Z}^d), \forall \{C_n\}_{n\in\mathbb{N}};\ C_n \uparrow B^c, \exists\, b_B > 0; \forall A \subset B, \forall n \in \mathbb{N},\ \mu([A,B]|B(C_n)) > b_B;$

(ii) the sequence $\{\mu([A,B]|B(C_n))\}_{n\in\mathbb{N}}$ is equicontinuous.

Proof:

1) From the definition of a Gibbs measure, we have:

$$\forall\, n \in \mathbb{N},\ \mu([A,B]|B(C_n)) > 0 \quad .$$

In addition, the function $U(B\cup .)-U(.)$ is continuous on $P(\mathbb{Z}^d)$ provided with the discrete topology, because we have, if $B=\{z_1,\dots,z_n\}$, $B_1=\{z_{n+1},\dots,z_p\}$, $B_2=\{z_{n+1},\dots,z_q\}$, with $p<q$:

$$|(U(B\cup B_1)-U(B_1))-(U(B\cup B_2)-U(B_2))| < |\sum_{i=p}^{q} a_i| \xrightarrow[B\uparrow\mathbb{Z}^d]{} 0 \quad .$$

Therefore, $\mu([A,B]|B(B^c))$ is continuous on $P(\mathbb{Z}^d)$, which is compact for the discrete topology; hence we have:

$$\exists\, a_B,\ a_B'>0;\ a_B < \mu([A,B]|B(B^c)) < a_B' \quad ,$$

and

$$\forall\, n \in \mathbb{N},\ 0<b_B = \frac{a_B}{2^{|B|}a_B'} < \mu([A,B]|B(B(C_n)) \quad .$$

The uniform continuity of $U(B\cup .)-U(.)$ on the compact $P(\mathbb{Z}^d)$ implies equicontinuity.

2) $\{\mu([A,B]|B(C_n))\}_{n\in\mathbb{N}}$ is a martingale, which converges μ-almost everywhere to $\mu([A,B]|B(B^c))$, when C_n increases up to B^c [12]; because the sequence is equicontinuous, this convergence is uniform; hence the limit is continuous and μ, having its conditional probabilities $\mu([A,B]|B(B^c))$ continuous and strictly positive, is a Gibbs measure [13].

Now we give an important characterization of the set G_V of all Gibbs measures having V as interaction potential.

Proposition 3.2

Let V be a uniform interaction potential, invariant for every translation of $\mathbb{Z}^d$. Then the following statements are equivalent:

(i) G_V is the set of the accumulation points, in the set $M(\mathbb{Z}^d)$ of all probabilities on $(P(\mathbb{Z}^d),B(\mathbb{Z}^d))$ provided with the weak topology, of the sequences:

$$\left\{\sum_{\upsilon\in M(\mathbb{Z}^d)} a_n(\upsilon)\Phi'_n(\upsilon)\right\}_{n\in\mathbb{N}} ,$$

where only a finite number of the a_n's is strictly positive (the others being equal to zero) and such that $\sum_{n\in\mathbb{N}} a_n = 1$, and where Φ'_n is the transposed of the operator Φ_n defined by a sequence $\{C_n\}_{n\in\mathbb{N}}$ such that: $C_n \uparrow \mathbb{Z}^d$ and, for every h continuous on $P(\mathbb{Z}^d)$ and $G\subset\mathbb{Z}^d$:

$$\Phi_n(h)(G) \overset{\mu\text{-a.s.}}{=} \sum_{A\subset C_n} \mu([A,C_n]|B(C_n^c))(G)h(A\cup(G\setminus C_n)) .$$

(ii) G_V is the set of all equilibrium states for the potential U associated with V.

(iii) G_V is the set of the accumulation points in $M(\mathbb{Z}^d)$ of the sequences $\{\mu_p\}_{p\in\mathbb{N}}$, where μ_p belongs to G_{V_p} and V_p is an interaction potential verifying:

$$\|V_p - V\| \xrightarrow[p\to\infty]{} 0$$

(in particular, we can take V_p with a range M_p as follows:

$$V_p = V.\mathbb{I}_{\{B\in F(\mathbb{Z}^d);\exists z\in\mathbb{Z}^d;B+z\subset M_p=M\cap K_p\}} ,$$

where $K_p = \{z\in\mathbb{Z}^d; |z|<p\}$.)

Proof:

(i) Let us show that every μ in G_V belongs to the proposed set; after [14], there exists a sequence of convex combinations:

$$\left\{\sum_{Y\subset C_n^c} a_n(Y)\mu(.|B(C_n^c))(Y)\right\}_{n\in\mathbb{N}} \quad \text{converging to } \mu .$$

In addition, we have:

$$\forall n\in\mathbb{N}, \forall B\subset\mathbb{Z}^d, \ \Phi'_n(\delta_B)|_{B(C_n)}(.) = \mu(.|B(C_n^c))(B)$$

and therefore

$$\left\{\sum_{Y\subset C_n^c} a_n(Y)\Phi'_n(\delta_Y)\right\}_{n\in\mathbb{N}} \quad \text{converges to } \mu .$$

Reciprocally, we can choose υ from the set of Dirac measures on $P(\mathbb{Z}^d)$; after [14], the accumulation points of sequences of convex combinations such as:

$$\left\{\sum_{Y\subset C_n^c} a_n(Y)\Phi'_n(\delta_Y)\right\}_{n\in\mathbb{N}} \quad \text{belongs to } G_V .$$

But these combinations are equal to $\Phi'_n\left(\sum_{Y \subset C_n^c} a_n(Y)\delta_Y\right)$ and the set of convex combinations of Dirac measures is weakly dense in $M(\mathbb{Z}^d)$; then, because Φ'_n is continuous, $M(\mathbb{Z}^d)$ weakly complete and G_V closed, we can replace the convex combinations by any probability υ.

(ii) Every μ in G_V is an equilibrium state for U [15].

Reciprocally, we have:

$$\forall\, B \in F(\mathbb{Z}^d),\quad \sum_{A \subset B} \mu([A,B]\,|\,B(B^c)) \overset{\mu\text{-a.s.}}{=} 1 \quad .$$

Therefore, if μ is an equilibrium state for U, we have:

$$\mu([A,B]\,|\,B(B^c))(Y) \overset{\mu\text{-a.s.}}{=} F_B(Y)\, e^{U(A\cup Y)-U(Y)}$$

and μ is a Gibbs measure for the interaction potential V associated with U.

(iii) The limit of every sequence $\{\mu_p\}_{p\in\mathbb{N}}$ belongs to G_V (cf.[5] or [16]).

Reciprocally, let μ be an element of G_V; after (i), if the sequence $\{C_n\}_{n\in\mathbb{N}}$ of elements in $F(\mathbb{Z}^d)$ is such that C_n increases up to $\mathbb{Z}^d$, there exists a sequence of convex combinations

$$\left\{\sum_{Y\subset C_n^c} a_n(Y)\left(\sum_{A\subset C_n} F_{C_n}(Y)\, e^{U(A\cup Y)-U(Y)}\, \delta_A\right)\right\}_{n\in\mathbb{N}} = \{\upsilon_n\}_{n\in\mathbb{N}} \quad \text{converging to } \mu \quad .$$

Because V is uniform, e^U is bounded below by a strictly positive number a such that:

$$a.2^{|C_n|} \leq \sum_{A\subset C_n} e^{U(A\cup Y)-U(Y)}$$

We are going to show that the sequences

$$\left\{\upsilon_{n,p} = \sum_{Y\subset C_n^c} a_n(Y)\left(\sum_{A\subset C_n} F_{C_n,n}(Y)\exp[U_p(A\cup Y)-U_p(Y)]\delta_A\right)\right\}_{p\in\mathbb{N}}$$

tend, uniformly in n, to υ_n (U_p being the potential associated with V_p). Now let d be the distance function for the metric space $M(\mathbb{Z}^d)$; then we have [17]:

$$d(\upsilon_n,\upsilon_{n,p}) \leq \inf\left(\eta\geq 0; \forall\{D_i\}_{i=1,k}\subset F(\mathbb{Z}^d),\ \forall A_i\subset D_i,\ \left|(\upsilon_n-\upsilon_{n,p})\left(\bigcup_{i=1}^{k}[A_i,D_i]\right)\right|\leq\eta\right) \quad .$$

Hence we have:

$$(A_p) \Rightarrow \forall\, n\in\mathbb{N},\ d(\upsilon_n,\upsilon_{n,p}) \leq \varepsilon \quad ,$$

where (A_p) is the following assertion:

$$\forall\, n \in \mathbb{N},\ \forall A \subset C_n,\ \left|\sum_{Y\subset C_n^c} a_n(Y)\left(e^{U(A\cup Y)} - e^{U_p(A\cup Y)}\right)\right| \leq \frac{\varepsilon}{2^{|C_n|}}\ .$$

But (A_p) is true, if the following assertion $(\mathcal{B}_p)$ is true:

$$\forall\, n \in \mathbb{N},\ \forall A \subset C_n,\ \forall Y \subset C_n^c,\ \left|e^{U(A\cup Y)} - e^{U_p(A\cup Y)}\right| \leq \frac{\varepsilon}{2^{|C_n|}}\ .$$

Let us show that $(\mathcal{B}_p)$ is true; we can set:

$$F_{C_n}(Y)\ e^{U(A\cup Y)} = \frac{k}{K_n} \quad \text{and} \quad F_{C_n,p}(Y)\ e^{U_p(A\cup Y)} = \frac{k_p}{K_{n,p}}\ .$$

Because $||V_p - V||$ tends to zero, we have:

$$\forall\ \varepsilon>0;\ a>\varepsilon,\ \exists\, p(\varepsilon) \in \mathbb{N},\ \forall p \geq p(\varepsilon),\ \forall n \in \mathbb{N},\ |k-k_p| \leq \varepsilon \quad \text{and} \quad |K_n - K_{n,p}| \leq 2^{|C_n|}\varepsilon\ .$$

Therefore, we have:

$$-\frac{\varepsilon}{K_n - 2^{|C_n|}\varepsilon}\left(1 + \frac{k2^{|C_n|}}{K_n}\right) \leq \frac{k}{K_n} - \frac{k_p}{K_{n,p}} \leq \frac{\varepsilon}{K_n + 2^{|C_n|}\varepsilon}\left(1 + \frac{k2^{|C_n|}}{K_n}\right)\ .$$

But $2^{|C_n|}a \leq K_n$; hence we have, if b is an upper bound for e^U: for every $p \geq p[\varepsilon(a+\varepsilon)/(1+b/a)]$, $(\mathcal{B}_p)$ is true, therefore (A_p) is true and $\{\upsilon_{n,p}\}_{p\in\mathbb{N}} \to \upsilon_n$ uniformly in n. Now we can construct a sequence of weak accumulation points of the sequences $\{\upsilon_{n,p}\}_{n\in\mathbb{N}}$ converging to μ.

Finally, we give a characterization of G_V by means of markovian fields.

Proposition 3.3

Let M be any subset of $\mathbb{Z}^d$; then we can associate with every homogeneous M-Markov field μ an interaction potential V such that μ belongs to G_V. In addition, if V is uniform, then V is translation invariant and has a range M.

Reciprocally, if V is uniform, translation invariant and having a range M, every μ in G_V is a homogeneous M-Markov field.

Proof:

1) After [13], the property (i) for μ implies that there exists an interaction potential V such that μ belongs to G_V.

Now we are going to show that V is translation invariant and has a range M, if V is uniform.

Let us denote by M_p the set $M \cap \{z \in \mathbb{Z}^d;\ |z| \leq p\}$; if we consider the interaction potential V_p defined as in proposition 3.2, we can show that:

$$||V_p - V|| \xrightarrow[p\to+\infty]{} 0\ ;$$

then, because V_p has a range M_p, V has a range M. In addition, if we prove that V_p is translation invariant, V is also translation invariant:

A) Let us prove that $\|V_p - V\| \xrightarrow[p\to+\infty]{} 0$. Because the subset $\mathbb{B}$ of uniform interaction potentials is closed in the set of all interaction potentials, there exists V' in $\mathbb{B}$, which is a uniform accumulation point of the V_p's, i.e., such that there exists a subsequence $\{V_n\}_{n\in\mathbb{N}} \subset \{V_p\}_{p\in\mathbb{N}}$ converging to V' in $\mathbb{B}$.

After proposition 3.2, if μ' belongs to $G_{V'}$, then there exists a sequence $\{\mu_n\}_{n\in\mathbb{N}}$ such that:

$$\forall n\in\mathbb{N},\ \mu_n \in G_{V_n} \quad \text{and} \quad \mu_n \xrightarrow[n\to+\infty]{} \mu' \quad .$$

Then we have, if

$$K_p^* = K_p \setminus \{0\}:\ \forall m\in\mathbb{N},\ \forall n\geq m,\ \forall G\subset\mathbb{Z}^d \quad ,$$

$$\mu_n([\{0\},\{0\}]|B(K_p^*))(G) \overset{\mu_n\text{-a.s.}}{=} F_{K_{p,n}^*}(G)\ \exp[U_n(\{0\}\cup(G\cap K_p^*))-U_n(G\cap K_p^*)]$$

$$= F'_{K_p^*}(G)\ \exp[U'(\{0\}\cup(G\cap K_p^*))-U'(G\cap K_p^*)] \overset{\mu\text{-a.s.}}{=} \mu([\{0\},\{0\}]|B(K_p^*))(G) \quad .$$

But we have also, because μ is M-markovian:

$$\mu([\{0\},\{0\}]|[G\cap K_p^*,K_p^*]) \overset{\mu\text{-a.s.}}{=} \mu([\{0\},\{0\}]|[G\cap\partial_M\{0\}\cap K_p^*,\partial_M\{0\}\cap K_p^*])$$

$$= \mu([\{0\},\{0\}]|[G\cap\partial_{M_p}\{0\}\cap K_p^*,\partial_{M_p}\{0\}\cap K_p^*])$$

$$\overset{\mu_n\text{-a.s.}}{=} F_{\partial M_p\{0\}}(G)\ \exp[U(\{0\}\cup(G\cap\partial_{M_p}\{0\}))-U(G\cap\partial_{M_p}\{0\})]$$

$$= F_{\partial M_p\{0\},n}(G)\ \exp[U_n(\{0\}\cup(G\cap\partial_{M_p}\{0\}))-U_n(G\cap\partial_{M_p}\{0\})]$$

$$\overset{\mu_n\text{-a.s.}}{=} \mu_n([\{0\},\{0\}]|B(K_p^*))(G) \quad ,$$

because μ_n is M_p-markovian [11].

Hence we have:

$$\forall G\subset\mathbb{Z}^d,\ F_{K_p^*}(G)\ \exp[U(\{0\}\cup(G\cap K_p^*))-U(G\cap K_p^*)]$$

$$= F'_{K_p^*}(G)\ \exp[U'(\{0\}\cup(G\cap K_p^*))-U'(G\cap K_p^*)] \quad .$$

In addition, we have: $V'(\{0\}) = V(\{0\})$, by construction of V'. Then these equalities imply, by induction on cardinality of B that:

$$\forall B\in F(\mathbb{Z}^d),\ V'(B) = V(B) \quad .$$

Consequently we have well:

$$||V_p - V|| \xrightarrow[p\to+\infty]{} 0 \quad .$$

B) The fact that μ is homogeneous implies that μ' is homogeneous; hence μ_n is homogeneous and, after [11], V_n is translation invariant.

2) If V is uniform, translation invariant, with a range M, there exists, after proposition 3.2, a sequence $\{\mu_p\}_{p\in\mathbb{N}}$ converging to μ; for every p, μ_p is a homogeneous M_p-Markov field [11], hence we have:

$$\forall\, B\subset\mathbb{Z}^d,\ \forall x\in B^c,\ \partial_M\{x\}\subset B \Rightarrow \partial_{M_p}\{x\}\subset B_p = B\cap K_p \quad .$$

Consequently, μ satisfies:

$$(jj)_p,\quad \mu_p([\{x\},\{x\}]\,|\,B(B_p)) \overset{\mu_p\text{-a.s.}}{=} \mu_p([\{x\},\{x\}]\,|\,B(\partial_{M_p}\{x\})) \quad .$$

We have seen in the proof of proposition 3.1 that, for every Gibbs state ν and for every B in $P(\mathbb{Z}^d)$, we have:

$$\nu(.\,|B(B_p)) \xrightarrow[p\to+\infty]{\nu\text{-a.s.}} \nu(.\,|B(B)) \quad .$$

Therefore, because μ_p tends to μ, the property $(jj)_p$, valid for every p, implies that the property (ii) is valid for μ. In the same way, we can prove that μ is homogeneous.

Finally, the property (i) for μ results from the fact that μ is in G_V. Now, we give a general result concerning the criticality of the interaction potential V.

Proposition 3.4

If V is a supermodular, translation invariant interaction potential with a range M belonging to $F(\mathbb{Z}^d)$, then V is critical (i.e., $|G_V|>1$) if we have:

$$\sum_{k=0}^{|\partial_M\{0\}|} (-1)^k \sum_{\substack{A\subset\partial_M\{0\}\\ |A|=k}} \mu([\{0\},\{0\}]\,|\,[A,\partial_M\{0\}]) = 0 \quad ,$$

where μ is any element of G_V.

Proof:

Let us denote by m(A) the conditional probability $\mu([\{0\},\{0\}]\,|\,[A,\partial_M\{0\}])$; the condition above expresses the fact that the linear equations of projectivity between the $\mu([A,\partial_M\{0\}])$'s are linearly dependent on the M-Markovian equation:

$$\sum_{A\subset\partial_M\{0\}} m(A)\mu([A,\partial_M\{0\}]) = \mu([\{0\},\{0\}]) \quad .$$

An equation of projectivity between $\mu([A,\partial_M\{0\}])$ and $\mu([A\backslash\{x\},\partial_M\{0\}])$, where x belongs to A, is for example:

$$\mu([A,\partial_M\{0\}]) + \mu([A\backslash\{x\},\partial_M\{0\}]) = \mu([A\backslash\{x\},(\partial_M\{0\})\backslash\{x\}]) \quad .$$

We can consider the $2^{|\partial_M\{0\}|}$th order matrix H defined by $(2^{|\partial_M\{0\}|}-1)$ linearly independent equations of projectivity and by the M-Markovian equation; then it suffices that the determinant of H is equal to zero [6].

Remark:

This result generalizes the classical theorem of D. RUELLE [18] or [19] for the case of pair interaction, but it is less strong, because we have only a sufficient condition of criticality.

Finally, we are going to show that the behaviour of a uniform interaction potential V can be predicted by the behaviour of the approximant potentials V_p's.

Proposition 3.5

If M belongs to $P(\mathbb{Z}^d)$ and V is uniform, supermodular and translation invariant, then, if all interaction potentials V_p's defined as in proposition 3.2 are non-critical, then V is non-critical.

Proof:

As in the proof of proposition 3.2, we can show that there exists a unique accumulation point μ belonging to G_V, for the sequence $\{\mu_p\}_{p\in\mathbb{N}}$, where μ_p belongs to G_{V_p}.

4. Estimation

4.1 The Statistical Structure

Let us suppose that we observe a sample $(X_1,\ldots,X_m)$ of a homogeneous M-Markovian field X (where $M\in F(\mathbb{Z}^d)$) only by the means of a window B_n (for example, B_n is a cube of $\mathbb{Z}^d$ and the length of whose sides is n); it is in the nature of the problem to make this observation for fixed values of the field on the frontier $\partial_M B_n$, that is conditional to the occupation of Y_n in $\partial_M B_n$ (we can express the fact that z is occupied, for example by a particle, a sick person,..., by writing: X(z)=1). Then the statistical structure corresponding to this observation is the following:

$$(P(\mathbb{Z}^d),B(\mathbb{Z}^d),\{\mu_V(.|[Y_n,\partial_M B_n])\}_{V\in\Theta})^m$$

where μ_V denotes the canonical measure of the field X having the interaction potential V and where: $\mathbb{R}^{|P(M)|} \supset \theta = \theta_0 \cup \theta_1$, θ_0 being the set of all non-critical interaction potentials V, whose range is M.

Now the statistical problem is the following: to estimate V from an appropriate sample of size m observed by the means of the window B_n.

4.2 Vertical Sampling

The estimation problem is simple, because the statistical structure above is, for every n, an exponential structure, whose associated canonical structure is the image of the initial structure by the following statistic:

$$\left(Y_n T_A^i\right)_{\substack{i=1,\ldots,m \\ A\subset B_n}}$$

defined by:

$$\forall\, C\subset B_n,\ Y_n T_A^i(C) = 1, \text{ if } C\supset A$$
$$= 0, \text{ otherwise} \quad .$$

Then a biased μ-a.s. convergent estimator of V(A) is given by:

$$\forall\, A\subset B_n,\ \overline{V}_m(A) = \sum_{B_n\supset C\supset A} (-1)^{|C\setminus A|} \log\left(\sum_{B_n\supset D\supset C} (-1)^{|B_n\setminus D|}\, Y_n \overline{T}_D^m\right) \quad ,$$

where $Y_n\overline{T}_D^m$ is the empirical mean of the $Y_n T_D^i$'s (for related cases, see also [1,2,3] and [20,21,22]).

4.3 Horizontal Sampling

In this case, we make only one observation of X by means of a wide window B_n, which satisfies: $B_n \uparrow \mathbb{Z}^d$. Then we must replace $Y_n\overline{T}_D^m$ by $Y_n S_D$ defined by:

$$\forall\, C\subset B_n,\ Y_n S_D(C) = \frac{|\{C\subset F\subset B_n; \exists z\in\mathbb{Z}^d; F=z+D\}|}{|\{F\subset B_n; \exists z\in\mathbb{Z}^d; F=z+D\}|} \quad .$$

4.4 Test to Measure the Range M

Because the range M is supposed to be known, we must have the possibility to test at least a necessary condition that the range is M; the M-Markovian character of X implies that we can use a test of sphericity of the normal asymptotic law [23] of statistics:

$$\left(\sqrt{m}\ {}_\phi\overline{T}^m_{\{x_i\}}\right)_{x_i\in B_n;\ \partial_M\{x_i\}\cap\partial_M\{x_j\}=\phi}$$

because ${}_\phi\overline{T}^m_{\{x_i\}}$ is an unbiased consistent estimator of $\mu([\{x_i\},\{x_i\}]\,|\,[\phi,B_n])$, where μ denotes the canonical measure of X. The M-Markovian character of X implies that the occupations of the x_i's are independent, conditionally to the occupation of the frontier $U\partial_M\{x_i\}$.

5. Properties of the Estimators

5.1 Asymptotic Properties

These properties are good only if $M = K_1 = \{z \in \mathbb{Z}^d; |z| \leq 1\}$, and if V is supermodular and non-critical; in fact, we have:

Proposition 5.1

Let us denote by H_{Y_n} the quantity $(\mathrm{Log}\, F_{B_n}(Y_n))/|B_n|$. Then, if V is supermodular and if the sequence $\{B_n\}_{n\in\mathbb{N}}$ tends to $\mathbb{Z}^d$, the following diagram holds:

$\forall n \in \mathbb{N}, \forall D \subset B_n,$

$$\begin{array}{ccc} Y_n^{\overline{T}^m_D}/|B_n| & \xrightarrow[m\to+\infty]{\mu\text{-p.s.}} & \dfrac{\partial H_{Y_n}}{\partial V(D)} \\ Y_n^{S_D}/|B_n| & & \Big\downarrow {\scriptstyle n\to+\infty} \\ & \searrow^{\mu\text{-p.s.}}_{n\to+\infty} & \dfrac{\partial H_\infty}{\partial V(D)} \end{array}$$

where H_∞ denotes the limit of H_{Y_n}, when B_n increases monotonically to $\mathbb{Z}^d$.

Proof:

This diagram results from the direct application of the convergence theorems given in [24,25,26] and from the classical properties of exponential structures [27].

Proposition 5.2

If $M = K_1$ and if V is non-critical ($V\in\theta_0$), then if $k_{n,D}$ denotes the cardinality of the set $\{A\subset B_n; \exists z\in\mathbb{Z}^d; z+A=D\}$, the following diagram holds:

$$\begin{array}{ccc} \left\{\sqrt{m}\; Y_n^{\overline{T}_D}/|B_n|\right\}_{D\subset B_n} & \xrightarrow[m\to+\infty]{L} & N(\overrightarrow{\mathrm{grad}}\; H_{Y_n}, D^2H_{Y_n}) \\ \left\{\sqrt{k_{n,D}}\; Y_n^{S_D}/|B_n|\right\}_{D\subset B_n} & & \Big\downarrow {\scriptstyle n\to+\infty}\; L \\ & \searrow^{L}_{n\to+\infty} & N(\overrightarrow{\mathrm{grad}}\; H_\infty, D^2H_\infty) \end{array}$$

where D^2H_∞ denotes the Hessian matrix of H_∞.

Proof:

This diagram results from the direct application of the central limit theorems given in [24,25,26].

5.2 Test of Criticality

Because the asymptotic properties depend on whether V is critical or not, we must have a test of criticality for V: in practice, we can only test a sufficient condition for criticality (which is also necessary if $M=K_1$ [18,19]): after proposition 3.4, if the vector $(\mu([\{0\},\{0\}]|[Y,\partial_M\{0\}]))_{Y\subset\partial_M\{0\}}$ belongs to a certain hyperplan of $\mathbb{R}^{2^{|\partial_M\{0\}|}}$, V is critical; therefore, we can use a normal test of mean equal to zero against mean different from zero, because the statistic:

$$\left(\sqrt{m}\ {}_{Y}\overline{T}^{m}_{\{0\}}\right)_{Y\subset\partial_M\{0\}}$$

is asymptotically normal, after proposition 5.2.

6. Random Ising Model

Let us suppose that each component of V is a discrete random variable, defined on a probability space (Ω,A,P) as follows:

$$\forall A\in F(\mathbb{Z}^d),\quad V(A) = v(A) \geq 0,\quad \text{with probability } p$$
$$= -v(A)\ ,\quad \text{with probability } 1-p\ .$$

Then the results concerning the exponential structure above still hold, if we convert the statistic ${}_{Y_n}T_A$ into the statistic ${}_{Y_n}T'_A$ defined as follows:

$${}_{Y_n}T'_A = {}_{Y_n}T_A\cdot sg(V(A))\ ,$$

where

$$sg(x) = +1\ ,\quad \text{if } x \geq 0$$
$$= -1\ ,\quad \text{if } x < 0\ .$$

The canonical exponential structure is the image by the statistic

$$\left({}_{Y_n}T'^{\,i}_A\right)_{\substack{i=1,\dots,m\\ A\subset B_n}}$$

of the new initial structure:

$$\left(\Omega \times P^{m}(\mathbb{Z}^{d}), A \otimes B^{\otimes m}(\mathbb{Z}^{d}), \{P \otimes \mu_{V}^{\otimes m}(\cdot \,|\, B(\partial_{M} B_{n}))\}_{V \in \Theta}\right) .$$

7. Conclusion

The search for more precise asymptotic properties for estimators requires new convergence theorems using a generalization of the classical inequalities for interaction potentials [28] and requires the knowledge of general results concerning inference on random fields [29]. This work is at present in progress.

Acknowledgements. I am very indebted to R. Coleman for reading and correcting my manuscript.

References

1. D.K. Pickard: J. Appl. Prob. *13*, 486 (1976)
2. D.K. Pickard: Adv. Appl. Prob. *9*, 476 (1977)
3. D.K. Pickard: J. Appl. Prob. *16*, 12 (1979)
4. J. Demongeot: Sèminaire de Statistiques de Grenoble *2*, 135 (1980)
5. J. Demongeot: Thesis, Grenoble (1975)
6. J. Demongeot: Commun. Math. Phys. (submitted)
7. K. Schürger, P. Tautu: Lecture Notes in Biomath. *11*, 92 (1976)
8. A.M. Wartell, E.W. Montroll: Adv. Chem. Phys. *22*, 129 (1972)
9. P.A. Vuillermot: Phys. Letters *61A*, 9 (1977)
10. C. Preston: Lecture Notes in Math. *534* (1976)
11. M.B. Averintsev: Theory of Prob. and its Appl. *17*, 20 (1972)
12. J. Neveu: *Martingales à temps discret*, (Masson, Paris 1972)
13. C. Preston: *Gibbs states on countable sets* (Cambridge Univ. Press 1974)
14. O.E. Lanford: Lecture Notes in Physics *20*, 1 (1973)
15. F. Ledrappier: Commun. Math. Phys. *33*, 119 (1973)
16. O.E. Lanford, D. Ruelle: Commun. Math. Phys. *13*, 194 (1969)
17. Yu.V. Prokhorov: Theory of Prob. and its Appl. *1*, 158 (1956)
18. D. Ruelle: Annals of Phys. *69*, 364 (1972)
19. F. Spitzer: Lecture Notes in Math. *390*, 114 (1974)
20. K. Krickeberg: *Ecole d'été de St-Flour X*, Lecture Notes in Math. (to appear)
21. E. Glötzl, B. Rauchenschwandtner, R. Takacs: Preprint Univ. Linz (1979)
22. J.N. Darroch, S.L. Lauritzen, T.P. Speed: Annals of Prob. *8*, 522 (1980)
23. T.W. Anderson: *An introduction to multivariate analysis* (J. Wiley, New York 1958)
24. C.C. Neaderhouser: Annals of Prob. *6*, 207 (1978)
25. J.L. Lebowitz: Commun. Math. Phys. *33*, 313 (1973)
26. G. Gallavotti, A. Martin-Löf: Nuovo Cimento *25*, 425 (1975)
27. J.R. Barra: *Notions fondamentales de statistique mathématique* (Dunod, Paris 1971)
28. J. Demongeot: Sèminaire de Statistiques de Grenoble (to appear)
29. X. Guyon: Preprint Univ. Orsay (1980)

List of Contributors

Argemi, J. 179
Aubry, S. 79

Bonomi, E. 225
Burgan, J.R. 37

Caboz, R. 81
Cambiaggio, E.E. 86
Clerc, R.L. 47
Couot, J. 54
Cuozzo, F.C. 86

Della Dora, J. 3
Demetrius, L. 233
Demongeot, J. 254
Derrida, B. 153
Duffy, D.M. 132

Feix, M.R. 37
Fijalkow, E. 37

Gillot, C. 54,57
Gillot, G. 57
Golês Chacc, E. 64
Gumowski, I. 54,71

Hanusse, P. 203
Hartmann, Ch. 47
Hilhorst, H.J. 97

Jullien, R. 166,171

Lacolle, B. 104
Liotard, D. 213
Lonke, A. 81
Lutton, J.L. 225

Manneville, P. 116
Maynard, R. 125
Mira, C. 54
Moraux, M.P. 37
Moussa, P. 159,171
Munier, A. 37

Ottavi, H. 143

Penot, J.P. 213
Penson, K.A. 166
Pfeuty, P. 166,171

Ramis, J.P. 12
Rammal, R. 125
Reboul, T. 26
Rivier, N. 132
Rossetto, B. 179
Roussenq, J. 143

de Sêze, L. 116

Thomann, J. 12
Thomas, R. 180

Uzelac, K. 166,171

Van Ham, P. 194
Van Leeuwen, C. 97
Vannimenus, J. 153
Veltman, B.P.Th. 97
Verhamme, A. 194
Vogelij, H.N.J. 97